T0222465

Springer Undergraduate Mathematics Series

SUMS Readings

SUMS Readings is a collection of books that provides students with opportunities to deepen understanding and broaden horizons. Aimed mainly at undergraduates, the series is intended for books that do not fit the classical textbook format, from leisurely-yet-rigorous introductions to topics of wide interest, to presentations of specialised topics that are not commonly taught. Its books may be read in parallel with undergraduate studies, as supplementary reading for specific courses, background reading for undergraduate projects, or out of sheer intellectual curiosity. The emphasis of the series is on novelty, accessibility and clarity of exposition, as well as self-study with easy-to-follow examples and solved exercises.

More information about this series at http://www.springer.com/series/16607

Charles H. C. Little • Kee L. Teo • Bruce van Brunt

An Introduction to Infinite Products

 Springer

Charles H. C. Little
Research Fellow and former Professor
of Mathematics Institute of
Fundamental Sciences
Massey University
Palmerston North, New Zealand

Kee L. Teo
Research Fellow and former Professor
of Mathematics Institute of
Fundamental Sciences
Massey University
Palmerston North, New Zealand

Bruce van Brunt
Associate Professor of Mathematics
Institute of Fundamental Sciences
Massey University
Palmerston North, New Zealand

ISSN 1615-2085   ISSN 2197-4144   (electronic)
Springer Undergraduate Mathematics Series
ISSN 2730-5813   ISSN 2730-5821   (electronic)
SUMS Readings
ISBN 978-3-030-90645-0   ISBN 978-3-030-90646-7   (eBook)
https://doi.org/10.1007/978-3-030-90646-7

Mathematics Subject Classification: 40A20, 26-01, 30-01

This Springer imprint is published by the registered company Springer Nature Switzerland AG
The registered company address is: Gewerbestrasse 11, 6330 Cham, Switzerland

Preface

Infinite series and products share a common mathematical heritage. A student usually gets a glimpse of series in a first course on the calculus followed by a substantial dose of the theory underlying infinite series in a course on elementary real analysis. In contrast, infinite products are usually relegated to a few comments and perhaps examples at this stage. If a student encounters infinite products, it is often in a second course on complex analysis in the context of representing functions given their distribution of zeros (e.g. Weierstrass factorization). This situation is understandable because series arguably play a larger rôle in analysis, and much of the theory for infinite products can be constructed from that for infinite series. Students might thus be forgiven if their exposure to products affirms that these creatures live in the shadow of infinite series. Indeed, this is often the case, yet a closer study shows that they have their own lives and manifest themselves through beautiful and unexpected relations.

The earliest algorithm for determining π dates back to Viète (c. 1590), who claimed that

$$\frac{2}{\pi} = \frac{\sqrt{2}}{2} \cdot \frac{\sqrt{2+\sqrt{2}}}{2} \cdot \frac{\sqrt{2+\sqrt{2+\sqrt{2}}}}{2} \cdots .$$

This is also the earliest published infinite product. Wallis (c. 1650) gave another expression for π:

$$\frac{\pi}{2} = \frac{2}{1}\frac{2}{3} \cdot \frac{4}{3}\frac{4}{5} \cdots \frac{4n^2}{4n^2-1} \cdots .$$

The theory underlying infinite processes (analysis) was young at this time, and no formal proof was available to support these claims.

The "father" of infinite products might be Euler (c. 1750). Among the cornucopia of results from his pen are the formulæ

$$\frac{\sin x}{x} = \left(\cos \frac{x}{2}\right)\left(\cos \frac{x}{2^2}\right)\left(\cos \frac{x}{2^3}\right)\cdots,$$

and

$$\frac{\sin x}{x} = \left(1 - \left(\frac{x}{\pi}\right)^2\right)\left(1 - \left(\frac{x}{2\pi}\right)^2\right)\left(1 - \left(\frac{x}{3\pi}\right)^2\right)\cdots.$$

Particular choices of x in these expressions lead to Viète's and Wallis's products. The alert student can immediately see relations between the Maclaurin series for $\sin x$ and Euler's products. Is it obvious that

$$\sin x = x\left(\left(1 - \left(\frac{x}{\pi}\right)^2\right)\left(1 - \left(\frac{x}{2\pi}\right)^2\right)\left(1 - \left(\frac{x}{3\pi}\right)^2\right)\cdots\right)$$

$$= x - \frac{x^3}{3!} + \frac{x^5}{5!} - \cdots ?$$

The interplay between infinite series and products is certainly interesting and Euler did not stop here. He went on to connect prime numbers with a certain series later to be called the Riemann zeta function:

$$\left(1 - \frac{1}{2^s}\right)^{-1}\left(1 - \frac{1}{3^s}\right)^{-1}\left(1 - \frac{1}{5^s}\right)^{-1}\cdots\left(1 - \frac{1}{p_j^s}\right)^{-1}\cdots$$

$$= 1 + \frac{1}{2^s} + \frac{1}{3^s} + \frac{1}{4^s} + \frac{1}{5^s} + \cdots,$$

where p_j is the jth prime number and $s > 1$. Riemann (c. 1860) took this result for a real test drive by considering s to be a complex variable. The Riemann zeta function proved pivotal in the theory of the distribution of prime numbers. It is still the starting point for a lot of mathematical research. Euler also derived a number of relations between products and series that form a basis for the study of partitions of integers—yet another current field of research.

Post Riemann, and independent of the aforementioned research directions, we note that von Seidel (c. 1871) showed that

$$\frac{\log x}{x - 1} = \frac{2}{1 + x^{1/2}} \cdot \frac{2}{1 + x^{1/2^2}} \cdot \frac{2}{1 + x^{1/2^3}} \cdots,$$

from which we get

$$\log 2 = \frac{2}{1 + \sqrt{2}} \cdot \frac{2}{1 + \sqrt{\sqrt{2}}} \cdot \frac{2}{1 + \sqrt{\sqrt{\sqrt{2}}}} \cdots.$$

Catalan (c. 1873) derived an infinite product expression for the number e, and some hundred years later Pippenger [55] derived the expression

$$\frac{e}{2} = \left(\frac{2}{1}\right)^{1/2} \left(\frac{2\,4}{3\,3}\right)^{1/2^2} \left(\frac{4\,6\,6\,8}{5\,5\,7\,7}\right)^{1/2^3} \cdots .$$

Infinite products continue to fascinate mathematicians.

This book is about the theory of infinite products and applications. The target readership is a student familiar with the basics of real analysis of a single variable and a first course in complex analysis up to and including the calculus of residues.

The first chapter is largely a summary of results on infinite series, with which the reader is assumed to be familiar. This chapter serves mostly to make the book more self-contained and it provides a platform to introduce notation and definitions. Except in the final section, there are no proofs or problems, and few examples. This chapter can be read as needed by the student. The final section covers double sequences and series. This material is seldom encountered in first courses on real or complex analysis. Here we provide proofs, examples and problems.

The second chapter contains the general theory of infinite products. Much of the material is standard in at least older books used for a second course on analysis, but we also include topics such as conditional convergence and Abel's theorem. The chapter finishes with Weierstrass and Blaschke products and factorization for analytic functions.

The remaining two chapters deal with applications of infinite products. The gamma and related functions (digamma, beta, etc.) are of great interest in their own right, but they also provide a nice application of the use of infinite products. We use infinite products to define the gamma function in Chap. 3, recover a number of well-known results and then consider functions related to the gamma function along with applications to determine the limit of certain types of infinite products.

Certainly, infinite products leave a sizable footprint in the distribution of prime numbers and also partition problems. The application of products to these topics in number theory is the subject of the final chapter. This is a huge field sprinkled with names like Euler, Jacobi, Riemann, Hardy and Ramanujan. We make no attempt to give a complete or even limited account of this topic. Nonetheless, we hope to whet the reader's appetite for further study.

The goal of this book is to provide the reader with a short introduction to infinite products and motivation for their study. Our aim is neither to provide a complete treatise on the subject nor an exhaustive compilation of results. The choice of results and applications is always debatable. There are clearly other avenues we might have pursued that are of equal importance. We hope to arm readers so that they might understand these products in the course of more advanced studies, and, on a different level, to inspire them with the sheer beauty of mathematics.

The support of Massey University is gratefully acknowledged. The book has certainly benefited from the reviewing process as the reviewers made a number of suggestions that improved the text. Finally, the authors appreciate the encouragement and support of their wives.

Palmerston North, New Zealand   Charles H. C. Little

Kee L. Teo

Bruce van Brunt

Contents

Chapter 1
Introduction

A number of results concerning infinite series are presented in this chapter. The material spans convergence tests for series of constants, uniform convergence for series of functions, and power series. It is not the intention here to give a comprehensive account of this theory, but rather to review material that is relevant to later chapters. This chapter is intended to make the rest of the book more self contained and remind the reader of some basic results. It is assumed that the reader has a background in real analysis of a single variable and elementary complex analysis; consequently, no attempt is made to prove these results or provide examples. Proofs and examples can be found in textbooks on real analysis such as [7, 23, 36] and complex analysis [2, 15, 56]. The exception is in the final section on double sequences and series. This material is often not part of courses on real or complex analysis. Here we provide proofs, examples and exercises.

This chapter can be used in two ways. It can be used as a brief review of infinite series to prime the study of the chapters that follow, or the reader can skip this chapter and return to it as might be necessary when reading the later chapters.

1.1 Series

Let $\{z_n\}$ be a sequence of real or complex numbers, and for each n let

$$S_n = \sum_{j=0}^{n} z_j.$$

The formal sum

$$z_0 + z_1 + z_2 + \cdots,$$

© The Author(s), under exclusive license to Springer Nature Switzerland AG 2022
C. H. C. Little et al., *An Introduction to Infinite Products*, SUMS Readings,
https://doi.org/10.1007/978-3-030-90646-7_1

is called an **infinite series**, or simply a **series**, and is denoted by

$$\sum_{j=0}^{\infty} z_j.$$

The number S_n is called the nth **partial sum** of the series and the numbers z_j are called the **terms** of the series. The series is called **real** if the terms are real and **complex** if the terms are complex. The series is said to **converge** if the sequence of partial sums $\{S_n\}$ converges; otherwise, it is said to **diverge**. If $\lim_{n\to\infty} S_n = S$ then we write

$$\sum_{j=0}^{\infty} z_j = S.$$

The number S is called the **sum** of the series.

Whether a series converges or diverges is the primary question in the theory of series. For a given series, however, it is seldom easy to determine S_n in a form conducive to evaluating a limit directly. A notable exception is the **geometric series**,

$$\sum_{j=0}^{\infty} z^j,$$

where z is a number. If $z = 0$ we interpret 0^0 as 1. For this series it can be shown that

$$S_n = \frac{1 - z^{n+1}}{1 - z},$$

provided $z \neq 1$, and

$$S_n = n + 1,$$

for $z = 1$. The definition of series convergence thus yields the following result.

Theorem 1.1.1 (Geometric Series) *If z is a complex number such that $|z| < 1$ then*

$$\sum_{j=0}^{\infty} z^j = \frac{1}{1 - z};$$

if $|z| \geq 1$ then the series diverges.

Another example is the telescoping series.

Theorem 1.1.2 (Telescoping Series) *The series*

$$\sum_{j=0}^{\infty}(z_{j+1} - z_j)$$

converges if and only if the sequence $\{z_n\}$ *converges, and in that case*

$$\sum_{j=0}^{\infty}(z_{j+1} - z_j) = \lim_{n\to\infty} z_n - z_0.$$

It is because S_n is difficult to procure in a form amenable to study convergence that indirect tests are required. These tests place restrictions on the terms (e.g. real, non-negative). We present some of these tests in the next section. The next result does not place such restrictions on the terms, but can be used to establish only that a series diverges.

Theorem 1.1.3 *If the series* $\sum_{j=0}^{\infty} z_j$ *converges, then* $\lim_{n\to\infty} z_n = 0$.

Corollary 1.1.4 (nth Term Test) *If* $\lim_{n\to\infty} z_n$ *does not exist or is nonzero, then the series* $\sum_{j=0}^{\infty} z_j$ *diverges.*

Finally we note that the question of convergence for a complex series can be reduced to that for real series.

Theorem 1.1.5 *A series* $\sum_{j=0}^{\infty} z_j$ *converges if and only if the series* $\sum_{j=0}^{\infty} \operatorname{Re}(z_j)$ *and* $\sum_{j=0}^{\infty} \operatorname{Im}(z_j)$ *both converge.*

The result above indicates that, in principle, convergence questions for complex series can be converted to convergence questions for real series. In practice, this result is less useful than it appears because the decomposition of the terms into real and imaginary parts can prove formidable.

1.2 Series with Non-Negative Terms

In this section we list a number of convergence tests for real series that have non-negative terms. Before we embark on these tests we note that the convergence or divergence of the series is not affected by any *finite* number of terms. We can always ignore a finite number of terms when testing a series for convergence. If, for instance, a result requires the terms a_j to be positive, we need only ensure that there is an integer N such that $a_j > 0$ for all $j > N$ and we ignore the first $N + 1$ terms. If the series converges then these terms affect only the value of the sum.

The first tests we present are called comparison tests. The idea is to use a series that is known to converge or diverge to deduce the convergence or divergence of another series.

Theorem 1.2.1 (Comparison Test) *Let $\{a_j\}$ and $\{b_j\}$ be sequences of non-negative numbers and suppose that there is an integer N such that*

$$a_n \leq b_n, \qquad (1.2.1)$$

for all $n \geq N$. If $\sum_{j=0}^{\infty} b_j$ converges then $\sum_{j=0}^{\infty} a_j$ converges; if $\sum_{j=0}^{\infty} a_j$ diverges then $\sum_{j=0}^{\infty} b_j$ diverges.

Inequalities such as (1.2.1) can sometimes be elusive (or tedious) to establish. In fact, it is the relative behaviour of a_n and b_n as $n \to \infty$ that is important. This relationship is reflected in the limit comparison test. Before we state this result, however, some definitions and notation need to be introduced.

Definition 1.2.1 Let $\{a_n\}$ and $\{b_n\}$ be sequences of positive terms. Then a_n and b_n are said to be of the **same order of magnitude** if there is a positive number L such that

$$\lim_{n \to \infty} \frac{a_n}{b_n} = L.$$

In this case we write $a_n \sim L b_n$. We say that a_n is of a **lesser order of magnitude** than b_n, and write $a_n \ll b_n$, if

$$\lim_{n \to \infty} \frac{a_n}{b_n} = 0.$$

Finally, a_n is of a **greater order of magnitude** than b_n if

$$\lim_{n \to \infty} \frac{a_n}{b_n} = \infty.$$

We then write $a_n \gg b_n$.

Theorem 1.2.2 (Limit Comparison Test) *Let $\sum_{j=0}^{\infty} a_j$ and $\sum_{j=0}^{\infty} b_j$ be series of positive terms.*

1. *If $a_n \sim L b_n$ for some $L > 0$, then both series converge or both diverge.*
2. *if $a_n \ll b_n$ and $\sum_{j=0}^{\infty} b_j$ converges, then so does $\sum_{j=0}^{\infty} a_j$.*
3. *If $a_n \gg b_n$ and $\sum_{j=0}^{\infty} b_j$ diverges, then so does $\sum_{j=0}^{\infty} a_j$.*

Comparison tests require the user to have a candidate series whose convergence or divergence is known. The user must thus have a library of series for which the behaviour is known. The real value of the comparison tests is that they can be used to exchange a series for one with simpler terms. The behaviour of the latter series might be deduced from some other test. The comparison series chosen should have terms that capture the same behaviour as the original series. The next result is useful in deciding what to use for a comparison series.

Theorem 1.2.3 *For any numbers $p > 0$ and $a > 1$,*

$$\log n \ll n^p \ll a^n \ll n! \ll n^n. \qquad (1.2.2)$$

Relation (1.2.2) contains functions commonly found in series terms. It is clearly not exhaustive. For instance $n \ll n \log n \ll n^2$ and $\log \log n \ll \log n$. Nonetheless, it gives the user a rough idea of what to select for a comparison series as the next example shows.

Example 1.2.1 Let

$$a_n = \frac{4^n - n^3}{(n+1)! + 2^n - \log(n+1)}.$$

Note first that a_n is positive for all large enough n, because $4^n \gg n^3$ and

$$(n+1)! \gg \log(n+1).$$

To select terms for a comparison series we identify the terms of greatest order in the numerator and the denominator. These terms are 4^n and $(n+1)!$ respectively. In this manner we arrive at a comparison series with terms

$$b_n = \frac{4^n}{(n+1)!},$$

and it follows that $a_n \sim b_n$. $\qquad\qquad \triangle$

The comparison test is also useful for devising series tests that depend only on the terms of the given series. For these tests the comparison series is built into the result. The next class of series tests are proved using the comparison test with the geometric series.

Theorem 1.2.4 (Ratio Test) *Let $\sum_{j=0}^{\infty} a_j$ be a series of positive terms and suppose there exist numbers r and N such that*

$$\frac{a_{n+1}}{a_n} \leq r < 1$$

for all $n \geq N$. Then the series converges. On the other hand, if

$$\frac{a_{n+1}}{a_n} \geq 1$$

for all $n \geq N$, then the series diverges.

Corollary 1.2.5 (Limit Ratio Test) *Let $\sum_{j=0}^{\infty} a_j$ be a series of positive terms and suppose*

$$\lim_{n \to \infty} \frac{a_{n+1}}{a_n} = L$$

for some number L. If L < 1 then the series converges; if L > 1 then the series diverges.

Theorem 1.2.6 (Root Test) *Let $\sum_{j=0}^{\infty} a_j$ be a series of non-negative terms and suppose that there exist numbers r and N > 0 such that $a_n^{1/n} \leq r < 1$ for all $n \geq N$. Then the series converges.*

Corollary 1.2.7 (Limit Root Test) *Let $\sum_{j=0}^{\infty} a_j$ be a series of non-negative terms and suppose*

$$\lim_{n \to \infty} a_n^{1/n} = L$$

for some number L. Then the series converges if L < 1 and diverges if L > 1.

Note that if $L = 1$ in the limit forms of the ratio or root test then the test fails in that no conclusion can be reached using these results. The ratio and root tests target series where the terms either decay as fast as q^n as $n \to \infty$ for some number q such that $0 < q < 1$, or grow at least as fast as Q^n as $n \to \infty$ for some number $Q > 1$. These tests are linked through the comparison test with the geometric series and it should occasion little surprise that if the ratio test fails because $L = 1$, the root test will also fail. Although these tests target series with the same type of behaviour, it is often easier to apply one than the other. Moreover, the root test requires only that the terms be non-negative and in this sense is more general. When these tests fail because $L = 1$ there are more delicate tests.

Theorem 1.2.8 (Kummer–Jensen Test) *Let $\sum_{j=0}^{\infty} a_j$ be a series of positive terms and $\{b_n\}$ a sequence of positive terms. Let*

$$c_n = b_n - \frac{a_{n+1}}{a_n} b_{n+1}$$

for all n.

1. *If $\lim_{n \to \infty} c_n > 0$ then $\sum_{j=0}^{\infty} a_j$ converges.*
2. *If $\sum_{j=0}^{\infty} 1/b_j$ diverges and there exists N such that $c_n \leq 0$ for all $n \geq N$, then $\sum_{j=0}^{\infty} a_j$ diverges.*

The test above is quite general but suffers the disadvantage of the user having to shop around for a suitable sequence $\{b_n\}$. If $b_n = 1$ then we get the limit form of the ratio test. If $b_n = n - 1$ then we get the following result.

Corollary 1.2.9 (Raabe's Test) *Let $\sum_{j=0}^{\infty} a_j$ be a series of positive terms and suppose that*

$$\lim_{n \to \infty} n \left(1 - \frac{a_{n+1}}{a_n} \right) = L$$

for some number L. Then the series converges if L > 1 and diverges if L < 1.

Yet another consequence of the Kummer-Jensen test is the following result.

Theorem 1.2.10 (Gauss's Test) *Let $\sum_{j=1}^{\infty} a_j$ be a series of positive terms. Suppose there exist a bounded sequence $\{s_n\}$ and a constant c such that*

$$\frac{a_{n+1}}{a_n} = 1 - \frac{c}{n} + \frac{s_n}{n^2} \tag{1.2.3}$$

for all $n > 0$. Then the series is convergent if $c > 1$ and divergent if $c \leq 1$.

The Kummer-Jensen test and Gauss's test require the user to identify a suitable sequence ($\{b_n\}$ or $\{s_n\}$) in order to apply the result. If we further restrict our conditions to non-negative terms that do not increase, then we have the following results.

Theorem 1.2.11 (Cauchy's Condensation Test) *If $\{a_n\}$ is a non-increasing sequence of non-negative terms, then the series $\sum_{j=0}^{\infty} a_j$ converges if and only if $\sum_{k=0}^{\infty} 2^k a_{2^k}$ does so.*

Cauchy's condensation test shows that the convergence or divergence of a given series is determined by a remarkably thin subsequence of its terms.

Theorem 1.2.12 (Integral Test) *Let f be a non-increasing continuous function on the interval $[1, \infty]$, and suppose that $f(x) \geq 0$ for all $x \geq 1$. Then the series $\sum_{j=1}^{\infty} f(j)$ and the integral $\int_1^{\infty} f$ both converge or both diverge.*

Series with terms such as $1/n$ and $1/n^2$ fail the ratio and root test because $L = 1$. In fact, Cauchy's condensation test and the integral test both show that a series with the former terms diverges whereas a series with the latter terms converges. The next result is useful when used in conjunction with the comparison test.

Theorem 1.2.13 (p-Series) *Let p be a real number. The series*

$$\sum_{n=1}^{\infty} \frac{1}{n^p}$$

converges if $p > 1$. If $p \leq 1$ then the series diverges.

1.3 Series with General Terms

The convergence tests of Sect. 1.2 require the series terms to be real and non-negative. *Prima facie*, this is a serious limitation. Clearly, if there are only a finite number of negative terms or all the a_n are negative, then the tests in the previous section can be applied. In this section we consider series where the terms are real but change sign infinitely many times, or the terms are complex.

Let $\{z_n\}$ be a sequence of real or complex numbers. The series $\sum_{j=0}^{\infty} z_j$ is said to **converge absolutely** if the series $\sum_{j=0}^{\infty} |z_j|$ converges. Note that the latter series has non-negative terms and the results of the previous section can be applied to this series to determine whether it converges. The value of absolute convergence is that it implies convergence.

Theorem 1.3.1 *If $\sum_{j=0}^{\infty} |z_j|$ converges, then so does $\sum_{j=0}^{\infty} z_j$.*

For a series with general terms z_j we can thus first test for absolute convergence. This leads to generalizations such as the following.

Theorem 1.3.2 (Generalized Ratio Test) *Suppose that*

$$\lim_{n \to \infty} \left| \frac{z_{n+1}}{z_n} \right| = L$$

for some number L. Then the series $\sum_{j=0}^{\infty} z_j$ converges absolutely if $L < 1$ but diverges if $L > 1$.

Theorem 1.3.3 (Generalized Root Test) *Suppose that*

$$\lim_{n \to \infty} |z_n|^{1/n} = L$$

for some number L. Then the series $\sum_{j=0}^{\infty} z_j$ converges absolutely if $L < 1$ but diverges if $L > 1$.

It is, of course, possible that a series converges but not absolutely. A series $\sum_{j=0}^{\infty} z_j$ is called **conditionally convergent** if the series converges but is not absolutely convergent. Roughly speaking, there are two mechanisms leading to the convergence:

1. the series converges because the terms $|z_j|$ decay quickly enough; and
2. the series converges because there are cancellations in the sum.

The distinction between absolutely and conditionally convergent series is important. The terms of an absolutely convergent series can be rearranged in any manner without changing the sum; in contrast, a conditionally convergent series depends on cancellations and thus on the order in which the terms are summed.

The terms of a convergent series may be grouped together by means of parentheses without affecting the convergence of the series. In other words,

$$\sum_{j=1}^{\infty} z_j = \sum_{k=0}^{\infty} \sum_{j=t_k+1}^{t_{k+1}} z_j, \tag{1.3.1}$$

where $\{t_n\}$ is any increasing sequence of positive integers such that $t_0 = 0$. On the other hand, if some terms are already grouped together, then removal of the parentheses used to group them may change a convergent series into a divergent one. Nevertheless we have the following theorem.

Theorem 1.3.4 *Suppose that $a_j \geq 0$ for each j. If the series*

$$\sum_{k=0}^{\infty} \sum_{j=t_k+1}^{t_{k+1}} a_j$$

converges to some number L, then so does $\sum_{j=1}^{\infty} a_j$.

If a series is conditionally convergent, then it is sensitive to regroupings. In addition, through rearrangements we can make this series sum to any number or diverge. Riemann's theorem summarizes this phenomenon.

Theorem 1.3.5 (Riemann) *For each conditionally convergent real series and given number S, the terms of the series may be rearranged to yield a series that converges to S. There is also a rearrangement of the terms so that the resulting series diverges.*

In contrast an absolutely convergent series is robust: it can be rearranged in any manner without changing the sum.

Theorem 1.3.6 (Dirichlet) *All rearrangements of an absolutely convergent series are absolutely convergent and converge to the same number.*

If a series $\sum_{j=0}^{\infty} z_j$ converges absolutely, then we know it is convergent; if $\sum_{j=0}^{\infty} |z_j|$ diverges, however, then the former series might converge conditionally. The convergence results presented so far cannot be applied in this case. We now give a few results that do not exclude absolute convergence, but include conditional convergence. Conditional convergence is rather sensitive and this is reflected in the assumptions made on the series terms. The first result concerns a class of real series known as **alternating series**, where the terms are of the form $(-1)^j b_j$, for a sequence of positive numbers $\{b_j\}$.

Theorem 1.3.7 (Leibniz's Test) *The alternating series $\sum_{j=0}^{\infty} (-1)^j b_j$ is convergent if $\{b_n\}$ is a non-increasing sequence of positive terms converging to 0.*

The extra structure of a convergent alternating series allows us to estimate the sum of the series directly in terms of b_j.

Theorem 1.3.8 *Let $\{b_n\}$ be a non-increasing sequence of positive terms and let*

$$\sum_{j=0}^{\infty} (-1)^j b_j = S.$$

Then

$$|S - S_n| \leq b_{n+1}$$

for each $n \geq 0$, where

$$S_n = \sum_{j=0}^{n} (-1)^j b_j.$$

Theorem 1.3.9 *Let $\{u_n\}$ and $\{v_n\}$ be sequences of complex numbers and suppose that:*

1. *$\sum_{j=0}^{\infty} |v_{j+1} - v_j|$ converges;*
2. *$\{v_n\}$ converges to 0; and*
3. *there is a constant K such that*

$$\left| \sum_{j=0}^{n} u_j \right| \leq K$$

for all $n \geq 0$.

Then $\sum_{j=0}^{\infty} u_j v_j$ is convergent.

Corollary 1.3.10 (Dirichlet's Test) *Let $\{u_n\}$ be a complex sequence and let $\{v_n\}$ be a real sequence. Suppose that:*

1. *$\{v_n\}$ is monotonic and converges to 0; and*
2. *there is a constant K such that*

$$\left| \sum_{j=0}^{n} u_j \right| \leq K$$

for all $n \geq 0$.

Then $\sum_{j=0}^{\infty} u_j v_j$ is convergent.

We close this section with a result concerning the multiplication of two series. The problem here is that multiplication opens the doors to a number of rearrangements, and given the above results this will be cause for concern. There is one clear case, however, when the series in question are absolutely convergent.

Theorem 1.3.11 *If $\sum_{j=0}^{\infty} a_j$ and $\sum_{j=0}^{\infty} b_j$ converge absolutely to A and B respectively, then*

$$\sum_{j=0}^{\infty} \sum_{k=0}^{j} a_k b_{j-k} = AB$$

and the convergence is absolute.

1.4 Uniform Convergence of Sequences and Series of Functions

The concept of convergence for sequences and series of functions is a natural extension of that of convergence of sequences and series of constants. Consider a sequence of functions $\{f_n\}$ defined on some subset Ω of $\mathbb{C}$. For each $z \in \Omega$ we can form a sequence of constants $\{f_n(z)\}$ and then consider the convergence of this sequence. Suppose that $\{f_n(z)\}$ converges for all $z \in \Omega$. Then $\{f_n\}$ defines a function f on Ω. A primary concern is whether the function f "inherits" properties such as continuity from $\{f_n\}$. For instance, if f_n is continuous on Ω for all n, is f continuous? A simple example shows that this is not, in general, true. Consider the (real) sequence defined by

$$f_n(x) = \frac{1}{1 + nx},$$

for $0 \leq x \leq 1$. Evidently f_n is continuous on this interval. For any x such that $0 < x \leq 1$ we have $f_n(x) \to 0$ as $n \to \infty$, but $f_n(0) = 1$. The function f is therefore not continuous at 0.

Roughly speaking, a convergent sequence of continuous functions need not converge to a continuous function, and similar statements can be made for integrable functions and differentiable functions. The essence of the problem is that there are two limits involved and the order in which these limits are taken is, in general, crucial. We thus seek conditions under which we can interchange the order of the limits and this search leads to the concept of uniform convergence.

Let $\{f_n\}$ be a sequence of functions defined on $\Omega \subseteq \mathbb{C}$. The sequence is said to **converge uniformly** to the function f on Ω if for every $\varepsilon > 0$ there exists an integer N such that

$$|f_n(z) - f(z)| < \varepsilon$$

for all $n \geq N$ and all $z \in \Omega$. Thus N may depend on ε but it must not depend on z. Throughout this section Ω will denote a nonempty subset of $\mathbb{C}$.

If $\{f_n\}$ is uniformly convergent to f on Ω, then $|f_n(z) - f(z)|$ necessarily has a supremum on Ω. We can therefore define

$$M_n = \sup_{z \in \Omega} |f_n(z) - f(z)|$$

for all n. The importance of the sequence $\{M_n\}$ is revealed in the following theorem.

Theorem 1.4.1 *The sequence $\{f_n\}$ converges uniformly to f on Ω if and only if the sequence $\{M_n\}$ converges to 0.*

Corollary 1.4.2 (Uniform Bound) *Suppose that $\{f_n\}$ converges uniformly to f on Ω. If M is a number such that $|f_n(z)| < M$ for all n and all $z \in \Omega$, then $|f(z)| \leq M$ for all $z \in \Omega$.*

A sequence $\{f_n\}$ of functions fails to be uniformly convergent to a function f on a set Ω if and only if there is an $\varepsilon > 0$ such that for all N there exist an integer $k \geq N$ and a $z_k \in \Omega$ such that

$$|f_k(z_k) - f(z_k)| \geq \varepsilon.$$

Uniform convergence satisfies the following analogue of the Cauchy principle.

Theorem 1.4.3 (Cauchy Principle) *Let $\{f_n\}$ be a sequence of functions defined on a set Ω. The sequence $\{f_n\}$ converges uniformly on Ω if and only if for each $\varepsilon > 0$ there is an integer N such that for all $z \in \Omega$ we have*

$$|f_n(z) - f_m(z)| < \varepsilon$$

whenever $m \geq N$ and $n \geq N$.

Uniform convergence on a set Ω may also be characterized in terms of sequences of numbers in Ω.

Theorem 1.4.4 *Let $\{f_n\}$ be a sequence of functions defined on Ω. Then the sequence converges uniformly to f on Ω if and only if*

$$\lim_{n \to \infty} (f_n(z_n) - f(z_n)) = 0$$

for each sequence $\{z_n\}$ in Ω.

Corollary 1.4.5 *If there exists a sequence $\{z_n\}$ in Ω such that*

$$\lim_{n \to \infty} (f_n(z_n) - f(z_n)) \neq 0,$$

then $\{f_n\}$ does not converge uniformly to f on Ω.

Theorem 1.4.6 *Let $\{f_n\}$ and $\{g_n\}$ be sequences that are uniformly convergent to f and g, respectively, on Ω.*

1. *The sequence $\{f_n + g_n\}$ is uniformly convergent to $f + g$ on Ω.*
2. *If f and g are bounded, then $\{f_n g_n\}$ is uniformly convergent to fg on Ω.*
3. *Suppose that f is bounded on Ω and there exists $M > 0$ such that $|g_n(z)| > M$ for all n and all $z \in \Omega$. Then $\{f_n/g_n\}$ is uniformly convergent to f/g on Ω.*

Part (2) of Theorem 1.4.6 implies the following corollary.

Corollary 1.4.7 *Let $\{f_n\}$ and $\{g_n\}$ be sequences that are uniformly convergent to f and g, respectively, on Ω. Suppose also that f_n and g_n are bounded for each n. Then $\{f_n g_n\}$ converges uniformly to fg on Ω.*

A function is **uniformly continuous** on Ω if for each $\varepsilon > 0$ there exists $\delta > 0$ such that

$$|f(z_1) - f(z_2)| < \varepsilon$$

whenever $\{z_1, z_2\} \subset \Omega$ and $|z_1 - z_2| < \delta$. Uniform continuity and uniform convergence can be combined to get the next result.

Theorem 1.4.8 *Suppose that $\{g_n\}$ is uniformly convergent to a function g on Ω, and let*

$$R_g = \{g_n(z) : n \in \mathbb{N}, z \in \Omega\} \cup \{g(z) : z \in \Omega\}.$$

Suppose f is a function that is uniformly continuous over a set J, where $R_g \subseteq J$. Then the sequence $\{f_n\}$ defined by

$$f_n(z) = f(g_n(z)),$$

for all $z \in \Omega$, is uniformly convergent on Ω to $f \circ g$.

We now review the pleasant properties of uniform convergence that justify the introduction of the concept.

Theorem 1.4.9 *Let $\{f_n\}$ be a sequence of functions that is uniformly convergent on Ω. Let c be an accumulation point of Ω, and suppose that $\lim_{z \to c} f_n(z)$ exists for all $n \in \mathbb{N}$. Then*

$$\lim_{n \to \infty} \lim_{z \to c} f_n(z)$$

exists if and only if

$$\lim_{z \to c} \lim_{n \to \infty} f_n(z)$$

exists, and in that case those limits are equal.

Corollary 1.4.10 *Let $\{f_n\}$ be a sequence of functions that is uniformly convergent to a function f on Ω. If f_n is continuous on Ω for each n, then f is continuous on Ω.*

There are similar results for uniformly convergent sequences of integrable functions and sequences of differentiable functions. The next four results are given for the special case of real-valued sequences of functions defined on an interval $I = [a, b] \subseteq \mathbb{R}$. These results can be extended to complex functions of a complex variable but they are partly eclipsed by results about uniformly convergent sequences of analytic functions. We will discuss results for analytic functions in the next section.

Theorem 1.4.11 *Let $\{f_n\}$ be a sequence of functions that are integrable over an interval I and suppose that $\{f_n\}$ converges uniformly to f on I. Then f is integrable over I and*

$$\lim_{n\to\infty} \int_a^b f_n = \int_a^b f. \tag{1.4.1}$$

Corollary 1.4.12 *Let $\{f_n\}$ be a sequence of functions that are continuous on the interval I and suppose that $\{f_n\}$ converges uniformly to f on I. Then f is integrable over I and Eq. (1.4.1) is satisfied.*

Corollary 1.4.13 *Let $\{f_n\}$ be a sequence of functions that are bounded and monotonic on the interval I and suppose that $\{f_n\}$ converges uniformly to f on I. Then f is integrable over I and Eq. (1.4.1) is satisfied.*

Theorem 1.4.14 *Let $\{f_n\}$ be a sequence of functions that are differentiable on the interval I. Suppose that the sequence $\{f_n'\}$ converges uniformly on I and that there exists $c \in I$ such that $\lim_{n\to\infty} f_n(c)$ exists. Then $\{f_n\}$ converges uniformly to a differentiable function f, and*

$$f'(x) = \lim_{n\to\infty} f_n'(x)$$

for all $x \in I$.

The concept of uniform convergence of a sequence of functions can readily be adapted to series of functions. Let $\{f_n\}$ be a sequence of functions defined on Ω and let $\{S_n\}$ be the sequence of partial sums defined by

$$S_n(z) = \sum_{j=0}^{n} f_j(z)$$

for all $z \in \Omega$. The series $\sum_{j=0}^{\infty} f_j(z)$ is said to **converge uniformly** on Ω if $\{S_n\}$ converges uniformly on Ω. Properties of uniformly convergent sequences can be used to derive analogous results for uniformly convergent series.

Theorem 1.4.15 *Let $\sum_{j=0}^{\infty} |f_j(z)|$ be a series that converges uniformly on Ω. Then the sequence $\{|f_n(z)|\}$ converges uniformly to 0 on Ω.*

Theorem 1.4.16 *Let $\sum_{j=0}^{\infty} |f_j(z)|$ be a series that converges uniformly on Ω. Then $\sum_{j=0}^{\infty} f_j(z)$ converges uniformly on Ω.*

Theorem 1.4.17 *Let $\sum_{j=0}^{\infty} f_j(z)$ be a series that converges uniformly on Ω. Let c be an accumulation point of Ω. Then*

$$\sum_{j=0}^{\infty} \lim_{z\to c} f_j(z) = \lim_{z\to c} \sum_{j=0}^{\infty} f_j(z),$$

provided one of the limits exists.

Corollary 1.4.18 *If $\sum_{j=0}^{\infty} f_j(z)$ converges uniformly on Ω and f_j is continuous on Ω for each j, then $\sum_{j=0}^{\infty} f_j(z)$ is continuous on Ω.*

The next two results we state for functions of a real variable defined on an interval $I = [a, b]$. We revisit the complex case in the next section.

Theorem 1.4.19 (Term-by-Term Integration) *Let $\{f_n\}$ be a sequence of functions that are integrable over the interval I. Suppose that the series $\sum_{j=0}^{\infty} f_j(x)$ is uniformly convergent on I. Then*

$$\sum_{j=0}^{\infty} \int_a^b f_j(x)\, dx = \int_a^b \sum_{j=0}^{\infty} f_j(x)\, dx.$$

Theorem 1.4.20 (Term-by-Term Differentiation) *Let $\{f_n\}$ be a sequence of functions that are differentiable on I. Suppose that the series $\sum_{j=0}^{\infty} f_j'(x)$ is uniformly convergent on I and that there exists $c \in I$ such that the series $\sum_{j=0}^{\infty} f_j(c)$ converges. Then the series $\sum_{j=0}^{\infty} f_j(x)$ is uniformly convergent on I to a differentiable function, and*

$$\left(\sum_{j=0}^{\infty} f_j(x) \right)' = \sum_{j=0}^{\infty} f_j'(x)$$

for all $x \in I$.

Tests for the uniform convergence of a series can be readily gleaned from convergence tests for series of constants. The key to such extensions is to ensure that inequalities are satisfied for all $z \in \Omega$. We list a sample of such extensions.

Theorem 1.4.21 (Comparison Test) *Let $\{f_n\}$ and $\{g_n\}$ be sequences of functions defined on Ω, and suppose that*

$$|f_n(z)| \le |g_n(z)|$$

for all $z \in \Omega$ and $n \in \mathbb{N}$. If the series $\sum_{j=0}^{\infty} |g_j(z)|$ is uniformly convergent on Ω, then so is $\sum_{j=0}^{\infty} |f_j(z)|$.

Corollary 1.4.22 (Weierstrass M-Test) *Let $\{f_n\}$ be a sequence of functions defined on Ω. Suppose there exists a sequence $\{M_n\}$ of constants such that $|f_n(z)| \le M_n$ for all $z \in \Omega$ and all n. If $\sum_{j=0}^{\infty} M_j$ converges, then $\sum_{j=0}^{\infty} f_j(z)$ converges uniformly on Ω to a bounded function.*

Theorem 1.4.23 (Ratio Test) *Let $\{f_n\}$ be a sequence of functions defined on Ω. Suppose there exist numbers $r < 1$ and N such that f_N is bounded and $f_n(z) \ne 0$ and*

$$\left| \frac{f_{n+1}(z)}{f_n(z)} \right| \le r$$

for all $z \in \Omega$ and $n \ge N$. Then the series $\sum_{j=0}^{\infty} f_j(z)$ converges uniformly on Ω.

Theorem 1.4.24 (Root Test) *Let $\{f_n\}$ be a sequence of functions defined on a set Ω. Suppose there exist numbers $r < 1$ and N such that*

$$|f_n(z)|^{1/n} \le r$$

for all $z \in \Omega$ and all $n \ge N$. Then the series $\sum_{j=0}^{\infty} f_j(z)$ converges uniformly on Ω.

Theorem 1.4.25 (Modified Dirichlet Test) *Let $\{u_n\}$ and $\{v_n\}$ be sequences of functions defined on Ω. Suppose that $\sum_{j=0}^{\infty} u_j(z)$ is uniformly convergent on Ω and that there is a constant K such that, for all $z \in \Omega$,*

$$\sum_{j=0}^{\infty} |v_{j+1}(z) - v_j(z)| \le K$$

and

$$|v_0(z)| \le K.$$

Then $\sum_{j=0}^{\infty} u_j(z) v_j(z)$ is uniformly convergent on Ω.

1.5 Analytic Functions and Power Series

The theory of analytic functions forms an important branch of analysis. We assume that the reader is familiar with the basics of complex analysis. In this section we introduce terms and notation, briefly discuss analytic functions, and then focus on power series. We first discuss sets in the complex plane.

Let $z_0 \in \mathbb{C}$. The **open disc** of **radius** r **centred** at z_0 is given by

$$D(z_0; r) = \{z \in \mathbb{C} : |z - z_0| < r\};$$

the **closed disc** by

$$\bar{D}(z_0; r) = \{z \in \mathbb{C} : |z - z_0| \le r\};$$

and the **punctured disc** by

$$D'(z_0; r) = \{z \in \mathbb{C} : 0 < |z - z_0| < r\}.$$

Thus an accumulation point of a subset Ω of $\mathbb{C}$ is a number z such that $D'(z; r) \cap \Omega \neq \emptyset$ for each $r > 0$. A set $\Omega \subseteq \mathbb{C}$ is **open** if for each $z \in \Omega$ there is a number $r > 0$ such that $D(z; r) \subset \Omega$. A set Ω is **connected** if Ω cannot be represented as a union of two disjoint nonempty sets. A **domain** is a nonempty open connected set, and a **region** is a union of a domain D and a subset of the set of accumulation points of D. A set Ω is **closed** if it contains all of its accumulation points. A set Ω is **bounded** if there is a number M such that $|z| \leq M$ for all $z \in \Omega$. Finally, if a set is both closed and bounded it is called **compact**.

Let f be a complex-valued function of a complex variable $z = x + iy$, where $x, y \in \mathbb{R}$. The function f can always be decomposed to the form

$$f(z) = u(x, y) + i v(x, y),$$

where u and v are real-valued functions. If u and v are continuous at (x_0, y_0) then f will be continuous at $z_0 = x_0 + iy_0$ and *vice versa*. At this level, continuity of f is just a problem in continuity of functions of two real variables. It is differentiation that distinguishes the analysis of functions of two real variables from that of complex functions. The student learns early that not every choice of functions u and v with smooth partial derivatives at z_0 leads to a function that is differentiable on an open disc containing z_0. The functions u and v must satisfy a coupled system of partial differential equations, (the Cauchy-Riemann equations). These equations are necessary, and f may still not be differentiable even if they are satisfied. Note that "complexifying" a differentiable function of a real variable will not, in general, lead to a differentiable complex function. For instance $f(x) = |x|$ is differentiable at all $x \in \mathbb{R} - \{0\}$, but $f(z) = |z|$ is not differentiable at any $z \in \mathbb{C}$.

It is possible that the complex function f is differentiable at an isolated point, i.e., f is differentiable at some point $z_0 \in \mathbb{C}$ but there is an $r > 0$ such that f is not differentiable anywhere in $D'(z_0; r)$. More generally, it is possible that f is differentiable for all z on some curve in the complex plane, but not off this curve. These functions are of limited interest in analysis. Functions that are differentiable at all points in at least some domain play the central rôle in complex analysis.

Definition 1.5.1 (Analytic Function) A function f is said to be **analytic** at $z_0 \in \mathbb{C}$ if there is a number $r > 0$ such that f is differentiable at all $z \in D(z_0; r)$.

The following points should be noted.

1. Analyticity is a much stronger condition than differentiability. A function can be differentiable at a point but not analytic at this point.
2. The set $D(z_0; r)$ is open. This implies that f must be analytic at all $z \in D(z_0; r)$.

A function f is called **analytic on a set** $\Omega \subseteq \mathbb{C}$ if f is analytic at each point in Ω. If f is analytic on $\mathbb{C}$ it is called an **entire** function. Analytic functions are also called **holomorphic** by many authors.

As noted in Sect. 1.4 the integration and differentiation of uniformly convergent sequences of functions can be generalized to accommodate complex functions. Integration of complex functions generally involves a contour in the complex plane.

Briefly, suppose that z is a complex function of a real variable t. If z is continuous on an interval $[a, b]$, then the set $\{z(t) : t \in [a, b]\}$ defines a **contour** C. If z is continuous and piecewise differentiable then the **contour integral** of f on C is

$$\int_C f(z)\, dz = \int_a^b f(z(t)) z'(t)\, dt.$$

A contour integral can always be decomposed into the form

$$\int_C f(z)\, dz = \int_a^b g(t)\, dt + i \int_a^b h(t)\, dt,$$

where g and h are real-valued functions. Let $\{f_n\}$ be a uniformly convergent sequence of functions integrable on C that converges to f. The decomposition above, along with Theorem 1.4.11, can be used to establish that f is also integrable on C. This result extends immediately to uniformly convergent series.

Theorem 1.4.14 can be generalized for complex functions, but the following result largely replaces this generalization.

Theorem 1.5.1 *Let $\{f_n\}$ be a sequence of functions analytic in a domain Ω. Suppose that f_n converges uniformly to f on every compact subset of Ω. Then f is analytic on Ω.*

A **power series** in a variable z is defined as a series of the form

$$\sum_{j=0}^{\infty} c_j (z - z_0)^j,$$

where $z_0, c_0, c_1, \ldots$ are constants. The number z_0 is called the **centre** of the series. The set of points for which a power series converges has a particularly simple geometry. Each power series has associated with it a **circle of convergence** in the interior of which the power series is absolutely convergent and in the exterior of which it diverges. The **radius of convergence** of the power series is that of the circle of convergence. We admit the possibilities that this radius may be 0 or infinite. In summary, we have the following theorem.

Theorem 1.5.2 *For the power series $\sum_{j=0}^{\infty} c_j (z - z_0)^j$ one of the following possibilities must hold:*

1. *the series converges only when $z = z_0$;*
2. *the series is absolutely convergent for all z;*
3. *for some $R > 0$ the series converges absolutely whenever $|z - z_0| < R$ and diverges whenever $|z - z_0| > R$.*

In the first case, we write $R = 0$; in the second case we write $R = \infty$.

The radius of convergence for a power series is clearly an important number. There are various formulæfor determining this number for a given power series. The general formula is given in the next result. Its proof is based on the root test.

Theorem 1.5.3 (Cauchy-Hadamard Formula) *Let $\sum_{j=0}^{\infty} c_j(z - z_0)^j$ be a power series with radius of convergence R.*

1. *If $\limsup_{n\to\infty} |c_n|^{1/n} = L$, where $0 < L < \infty$, then $R = 1/L$;*
2. *If $\limsup_{n\to\infty} |c_n|^{1/n} = 0$, then $R = \infty$;*
3. *If $\limsup_{n\to\infty} |c_n|^{1/n} = \infty$, then $R = 0$.*

The Cauchy-Hadamard formula can prove awkward to use if we must deal with limits superior. If, however, $\lim_{n\to\infty} |c_n|^{1/n}$ exists or is infinite then the formula is more convenient because

$$\limsup_{n\to\infty} |c_n|^{1/n} = \lim_{n\to\infty} |c_n|^{1/n}.$$

The ratio test can also be used to determine R in certain cases. In particular we have the following result.

Theorem 1.5.4 *Let $\sum_{j=0}^{\infty} c_j(z - z_0)^j$ be a power series and suppose that*

$$\lim_{n\to\infty} \left| \frac{c_{n+1}}{c_n} \right| = L.$$

If $0 < L < \infty$, then $R = 1/L$; if $L = 0$, then $R = \infty$.

If the radius of convergence is positive, Theorem 1.5.2 shows that the series converges in the set $D(z_0; R)$ and thereby defines a function f in this set. The set $D(z_0; R)$ is called the **disc of convergence** (if $R = \infty$ this set is $\mathbb{C}$). In fact, the series converges uniformly in any compact subset of $D(z_0; R)$. Theorem 1.5.1 thus yields the next result.

Theorem 1.5.5 *Let $\sum_{j=0}^{\infty} c_j(z - z_0)^j$ be a power series with a radius of convergence $R > 0$. Then the function f defined on $D(z_0; R)$ by this power series is analytic. Moreover,*

$$f'(z) = \sum_{j=1}^{\infty} jc_j(z - z_0)^{j-1},$$

for all $z \in D(z_0; R)$.

In the theorem above, note that f' is also defined by a power series. It can be shown that the radius of convergence for this series is the same as that for f. The theorem can thus be applied to the power series for f'. In fact we can apply this argument any number of times. Evidently, f has derivatives of all orders in $D(z_0; R)$: differentiation preserves analyticity within the disc of convergence.

If f is analytic at z_0, then it can be shown that f has derivatives of all orders. The connection between power series and analytic functions runs much deeper as the following result shows.

Theorem 1.5.6 (Taylor) *Suppose that f is analytic at z_0. Then there exist $r > 0$ and unique constants $c_0, c_1, \ldots$ such that*

$$f(z) = \sum_{j=0}^{\infty} c_j (z - z_0)^j,$$

for all $z \in D(z_0; r)$. The constants c_j are given by

$$c_j = \frac{f^{(j)}(z_0)}{j!}.$$

We thus see that f is analytic at z_0 if and only if f can be represented near z_0 by a power series centred at z_0 with a positive radius of convergence. The series given in this theorem is known as a **Taylor series** for f. In the case where $z_0 = 0$, it is also known as the **Maclaurin series** for f. Table 1.1 lists some useful Maclaurin series.

If we apply Taylor's theorem to a real-valued function f of a real variable x, then we may wish to approximate $f(x)$ by truncating a Taylor series. The following theorem is occasionally helpful in estimating the error incurred by doing so.

Theorem 1.5.7 *Let x_0 be a real number, n a non-negative integer and f a function such that $f^{(n+1)}(x)$ exists for all x in some open interval I containing x_0. Then for each $x \in I - \{x_0\}$ there exists a number ξ between x and x_0 such that*

$$f(x) = \sum_{j=0}^{n} \frac{f^{(j)}(x_0)}{j!} (x - x_0)^j + \frac{f^{(n+1)}(\xi)}{(n+1)!} (x - x_0)^{n+1}.$$

Suppose that the radius of convergence R for a power series centred at z_0 is positive and finite. We know that the series defines a function f that is analytic in $D(z_0; R)$, but it may be that f can be continued analytically beyond $D(z_0; R)$. We eschew a discussion about analytic continuation, but simply note that the power

Table 1.1 Some Maclaurin series

$e^z = \sum_{j=0}^{\infty} \frac{z^j}{j!}$	for all $z \in \mathbb{C}$
$\log(1+z) = \sum_{j=0}^{\infty} (-1)^j \frac{z^{j+1}}{j+1}$	for all z such that $|z| < 1$
$\sin z = \sum_{j=0}^{\infty} (-1)^j \frac{z^{2j+1}}{(2j+1)!}$	for all $z \in \mathbb{C}$
$\cos z = \sum_{j=0}^{\infty} (-1)^j \frac{z^{2j}}{(2j)!}$	for all $z \in \mathbb{C}$
$\arctan z = \sum_{j=0}^{\infty} (-1)^j \frac{z^{2j+1}}{2j+1}$	for all z such that $|z| < 1$

series for f can sometimes be used to define an analytic function that is the same as f in $D(z_0; R)$ but also analytic at points outside $D(z_0; R)$. For example, the geometric series of Theorem 1.1.1 defines a function analytic in $D(0; 1)$. This function corresponds to $1/(1 - z)$ which, in fact, is analytic for all $z \neq 1$.

A complex number w is a **singularity** of a function f if there exists a sequence $\{w_n\}$ such that $\lim_{n \to \infty} w_n = w$ and f is analytic at each w_n but not at w. Given a function f that is analytic at z_0, a useful observation is that if f is not entire then the radius of convergence for the power series representing f centred at z_0 must be finite and correspond to the distance from the closest singularities to z_0. In other words, if R is finite, then the circle of convergence

$$C(z_0; R) = \{z : |z - z_0| = R\}$$

contains at least one singularity of f. For instance, if $z_0 = 0$ and $f(z) = 1/(1 - z)$, then the closest singularity of f to 0 is at $z = 1$, and we see that the radius of convergence for the power series (the geometric series in this case) is 1. An extreme example is given by the power series $\sum_{n=0}^{\infty} z^{n!}$, which has radius of convergence 1 and defines a function having singularities at every point on $C(0; 1)$ (cf. [67, p. 159]).

Theorem 1.5.2 does not give any information about the convergence of a power series on $C(z_0; R)$. It is easy to construct examples where the series converges at all, some, or none of the points on this circle. Although the radius of convergence of a power series is unaffected by differentiation, the differentiated series will in general have different convergence behaviour on $C(z_0; R)$.

We finish this section with two results concerning the behaviour of a function represented by a power series as the variable approaches the circle of convergence. We state these results for real series, noting that they can be generalized to complex power series.

Theorem 1.5.8 (Abel) *Let f be a function that is represented by the power series*

$$f(x) = \sum_{j=0}^{\infty} a_j x^j,$$

and suppose that this series has a unit radius of convergence. Suppose further that the series $\sum_{j=0}^{\infty} a_j$ converges. Then $\sum_{j=0}^{\infty} a_j x^j$ is uniformly convergent on $I = [0, 1]$ and

$$\lim_{x \to 1^-} f(x) = \sum_{j=0}^{\infty} a_j.$$

The direct converse to Abel's theorem would assert that if $\lim_{x \to 1^-} f(x) = S$ then $\sum_{j=0}^{\infty} a_j$ converges to S. That this assertion is not true is illustrated by the geometric series

$$f(x) = \frac{1}{1+x} = \sum_{j=0}^{\infty} (-1)^j x^j,$$

for which $\lim_{x \to 1^-} f(x) = 1/2$, but $\sum_{j=0}^{\infty} (-1)^j$ diverges. A partial converse, however, can be salvaged if conditions are placed on the order of magnitude of a_j.

Theorem 1.5.9 (Tauber) *If $|a_n| \ll 1/n$ and*

$$\lim_{x \to 1^-} \sum_{j=0}^{\infty} a_j x^j = S,$$

then $\sum_{j=0}^{\infty} a_j$ converges to S.

1.6 Double Series

We now introduce the concept of a double series. As this idea might not be familiar to readers, we include proofs and examples. We also show how to use double series to prove Dirichlet's theorem (Theorem 1.3.6).

Let $\{z_{m,n}\}$ be a double sequence and z a complex number. We say that $\{z_{m,n}\}$ **converges** to z if for every $\varepsilon > 0$ there exists N such that $|z_{m,n} - z| < \varepsilon$ whenever $m \geq N$ and $n \geq N$. In this case we write

$$\lim_{(m,n) \to \infty} z_{m,n} = z.$$

We can arrange the terms of a double sequence $\{z_{m,n}\}$ in an infinite array:

$$
\begin{array}{llll}
z_{0,0} & z_{0,1} & z_{0,2} & \cdots \\
z_{1,0} & z_{1,1} & z_{1,2} & \cdots \\
\vdots & \vdots & \vdots & \ddots
\end{array}
\tag{1.6.1}
$$

For all non-negative integers m and n, we denote by $S_{m,n}$ the sum of the entries of the $(m+1) \times (n+1)$ leading rectangular subarray:

$$S_{m,n} = \sum_{j=0}^{m} \sum_{k=0}^{n} z_{j,k}.$$

We call $S_{m,n}$ a **partial sum** of the double sequence. If $z_{j,k} = 0$ whenever $j > 0$, then $S_{m,n}$ is just a partial sum of the sequence given by the first row of the array. Thus the theory developed in this section can be applied to ordinary series.

We write

$$\sum_{j,k=0}^{\infty} z_{j,k} = \lim_{(m,n)\to\infty} S_{m,n} \tag{1.6.2}$$

if the limit exists. We call

$$\sum_{j,k=0}^{\infty} z_{j,k}$$

a **double series** and say that it **converges** to S if Eq. (1.6.2) holds. It **diverges** if there is no S to which it converges.

The following result is analogous to one for ordinary series and can be proved in a similar manner.

Theorem 1.6.1 *If $\sum_{j,k=0}^{\infty} w_{j,k}$ and $\sum_{j,k=0}^{\infty} z_{j,k}$ are convergent, then*

$$\sum_{j,k=0}^{\infty} (s w_{j,k} + t z_{j,k}) = s \sum_{j,k=0}^{\infty} w_{j,k} + t \sum_{j,k=0}^{\infty} z_{j,k}$$

for any s and t.

There are various ways of adding the entries of the infinite array (1.6.1). For instance, we may add row by row, thereby obtaining

$$\sum_{j=0}^{\infty} \sum_{k=0}^{\infty} z_{j,k} = \sum_{j=0}^{\infty} \left(\sum_{k=0}^{\infty} z_{j,k} \right). \tag{1.6.3}$$

Column by column addition yields

$$\sum_{k=0}^{\infty} \sum_{j=0}^{\infty} z_{j,k} = \sum_{k=0}^{\infty} \left(\sum_{j=0}^{\infty} z_{j,k} \right). \tag{1.6.4}$$

The expressions given in Eqs. (1.6.3) and (1.6.4) are known as **iterated series**.

On the other hand, we might add up the entries of the leading square subarrays, giving partial sums of the form $S_{n,n}$ where n is a non-negative integer. We illustrate the calculation of $S_{2,2}$ by this method in Table 1.2.

Another alternative is addition diagonal by diagonal, giving partial sums of the form Δ_n where Δ_n, for any non-negative integer n, is the sum of the entries in the first $n + 1$ diagonals in the top left part of the array. The calculation of Δ_2 is illustrated in Table 1.3.

Table 1.2 $S_{2,2} = z_{0,0} + z_{0,1} + z_{1,1} + z_{1,0} + z_{0,2} + z_{1,2} + z_{2,2} + z_{2,1} + z_{2,0}$

$z_{0,0}$	$z_{0,1}$	$z_{0,2}\cdots$	
	$\downarrow$	$\downarrow$	
$z_{1,0} \leftarrow z_{1,1}$		$z_{1,2}\cdots$	
		$\downarrow$	
$z_{2,0} \leftarrow z_{2,1} \leftarrow z_{2,2}\cdots$			
$\vdots$	$\vdots$	$\vdots$	$\ddots$

Table 1.3 $\Delta_2 = z_{0,0} + z_{0,1} + z_{1,0} + z_{0,2} + z_{1,1} + z_{2,0}$

$z_{0,0}$	$z_{0,1}$	$z_{0,2}\cdots$	
	$\swarrow$	$\swarrow$	
$z_{1,0}$	$z_{1,1}$	$z_{1,2}\cdots$	
	$\swarrow$		
$z_{2,0}$	$z_{2,1}$	$z_{2,2}\cdots$	
$\vdots$	$\vdots$	$\vdots$	$\ddots$

In both of the last two approaches we are ordering the terms of the double series to form first a sequence $\{c_n\}$ and then the series $\sum_{j=0}^{\infty} c_j$. Note that this statement is not true for the two iterated series, however. More precisely, a sequence $\{c_n\}$ is called an **arrangement** of a double sequence $\{z_{m,n}\}$ if for each ordered pair (m, n) of non-negative integers there exists a unique non-negative integer j such that $c_j = z_{m,n}$ and for each non-negative integer j there is a unique ordered pair (m, n) of non-negative integers such that $c_j = z_{m,n}$. Each rearrangement of an ordinary sequence is a special case. With this terminology, the terms of the series constructed in each of the last two approaches are ordered by arrangements of the original double sequence.

Intuitively, a double series represents an attempt to add the terms in an infinite array, and we have described a number of methods for the formation of such a sum. It is natural to ask whether the convergence of the series is affected by the method chosen for the addition. Indeed it may be, but not if the terms in the array are non-negative (real) numbers. Let us first try some examples.

Example 1.6.1 For all $m \geq 0$ and $n \geq 0$ let

$$a_{m,n} = \begin{cases} 1 & \text{if } m = n + 1 \\ -1 & \text{if } m = n - 1 \\ 0 & \text{otherwise.} \end{cases}$$

The terms of the resulting double sequence are set forth in the table below.

$$
\begin{array}{rrrrrr}
0 & -1 & 0 & 0 & 0 & \cdots \\
1 & 0 & -1 & 0 & 0 & \cdots \\
0 & 1 & 0 & -1 & 0 & \cdots \\
0 & 0 & 1 & 0 & -1 & \cdots \\
\vdots & \vdots & \vdots & \vdots & \vdots & \ddots
\end{array}
$$

Here each $S_{m,n}$ is one of $0, 1, -1$ and the limit (1.6.2) does not exist. Moreover $\sum_{k=0}^{\infty} a_{0,k} = -1$ and $\sum_{k=0}^{\infty} a_{j,k} = 0$ for all $j > 0$, so that

$$
\sum_{j=0}^{\infty} \sum_{k=0}^{\infty} a_{j,k} = -1.
$$

Similarly

$$
\sum_{k=0}^{\infty} \sum_{j=0}^{\infty} a_{j,k} = 1,
$$

and so

$$
\sum_{j=0}^{\infty} \sum_{k=0}^{\infty} a_{j,k} \neq \sum_{k=0}^{\infty} \sum_{j=0}^{\infty} a_{j,k}.
$$

Note also that $S_{n,n} = \Delta_n = 0$ for all n. $\triangle$

Example 1.6.2 For all $m \geq 0$ and $n \geq 0$ define

$$
a_{m,n} = \begin{cases}
(-1)^n & \text{if } m = 0 \text{ and } n > 0 \\
(-1)^{m+1} & \text{if } m > 0 \text{ and } n = 0 \\
0 & \text{otherwise.}
\end{cases}
$$

The array of terms of this double sequence is given in the table below.

$$
\begin{array}{rrrrrr}
0 & -1 & 1 & -1 & 1 & \cdots \\
1 & 0 & 0 & 0 & 0 & \cdots \\
-1 & 0 & 0 & 0 & 0 & \cdots \\
1 & 0 & 0 & 0 & 0 & \cdots \\
-1 & 0 & 0 & 0 & 0 & \cdots \\
\vdots & \vdots & \vdots & \vdots & \vdots & \ddots
\end{array}
$$

Again the sequence $\{S_{m,n}\}$ diverges. Here $\sum_{k=0}^{\infty} a_{0,k}$ and $\sum_{j=0}^{\infty} a_{j,0}$ also diverge but $S_{n,n} = \Delta_n = 0$ for all n. $\triangle$

The next example is due to Brock [13].

Example 1.6.3 Let $\sum_{k=0}^{\infty} b_k = 0$, where $b_k \neq 0$ for all k, and define

$$a_{m,n} = (-1)^n b_m + (-1)^m b_n$$

for all non-negative m and n. For all odd n we have

$$\sum_{j=0}^{m} \sum_{k=0}^{n} (-1)^k b_j = \sum_{j=0}^{m} b_j \sum_{k=0}^{n} (-1)^k = 0.$$

Similarly if n is even, then

$$\sum_{j=0}^{m} \sum_{k=0}^{n} (-1)^k b_j = \sum_{j=0}^{m} b_j \to 0$$

as $m \to \infty$. Hence

$$\lim_{(m,n) \to \infty} \sum_{j=0}^{m} \sum_{k=0}^{n} (-1)^k b_j = 0.$$

Similarly

$$\lim_{(m,n) \to \infty} \sum_{j=0}^{m} \sum_{k=0}^{n} (-1)^j b_k = 0.$$

We conclude that the double sequence $\{S_{m,n}\}$ converges to 0. On the other hand, for each fixed j we have

$$\sum_{k=0}^{n} a_{j,k} = b_j \sum_{k=0}^{n} (-1)^k + (-1)^j \sum_{k=0}^{n} b_k.$$

Hence $\sum_{k=0}^{\infty} a_{j,k}$ diverges, since $b_j \neq 0$. Similarly $\sum_{j=0}^{\infty} a_{j,k}$ diverges for each k. Therefore neither

$$\sum_{j=0}^{\infty} \sum_{k=0}^{\infty} a_{j,k}$$

nor

$$\sum_{k=0}^{\infty} \sum_{j=0}^{\infty} a_{j,k}$$

exists. △

The next theorem is analogous to the Cauchy principle.

Theorem 1.6.2 *The double series $\sum_{j,k=0}^{\infty} z_{j,k}$ converges if and only if for each $\varepsilon > 0$ there exists N such that*

$$|S_{p,q} - S_{m,n}| < \varepsilon$$

whenever $p \geq m \geq N$ and $q \geq n \geq N$.

Proof The necessity of the condition is proved in the same way as for ordinary series, and so we shall prove only sufficiency.

Suppose the condition holds. Then given $\varepsilon > 0$ there exists N for which

$$|S_{p,q} - S_{m,m}| < \varepsilon$$

whenever $p \geq m \geq N$ and $q \geq m$. In particular,

$$|S_{p,p} - S_{m,m}| < \varepsilon.$$

By the Cauchy principle, $\{S_{m,m}\}$ converges to some number S. Hence there exists M such that

$$|S_{m,m} - S| < \varepsilon$$

for all $m \geq M$. Let $N_1 = \max\{M, N\}$. Then

$$|S_{p,q} - S| \leq |S_{p,q} - S_{m,m}| + |S_{m,m} - S| < 2\varepsilon$$

whenever $p \geq m \geq N_1$ and $q \geq m$. We conclude that

$$\sum_{j,k=0}^{\infty} z_{j,k} = S.$$

□

Theorem 1.6.3 *If the double series $\sum_{j,k=0}^{\infty} z_{j,k}$ converges, then*

$$\lim_{(m,n)\to\infty} z_{m,n} = 0.$$

Proof The theorem follows from the equation

$$z_{m,n} = S_{m,n} - S_{m-1,n} - S_{m,n-1} + S_{m-1,n-1},$$

which holds whenever $m > 0$ and $n > 0$, and the fact that

$$\lim_{(m,n)\to\infty} S_{m,n} = S$$

if $\sum_{j,k=0}^{\infty} z_{j,k}$ converges to S. $\square$

A double series $\sum_{j,k=0}^{\infty} z_{j,k}$ is **absolutely convergent** if $\sum_{j,k=0}^{\infty} |z_{j,k}|$ is convergent. We shall see that absolutely convergent double series are much better behaved than the examples we have considered. In fact, all the sums we have discussed converge to the same limit if the double series in question is absolutely convergent and any one of those sums converges. First we note that, as for ordinary series, absolutely convergent double series are indeed convergent. We state this result formally as a theorem.

Theorem 1.6.4 *If the double series*

$$\sum_{j,k=0}^{\infty} |z_{j,k}|$$

converges, then so does

$$\sum_{j,k=0}^{\infty} z_{j,k}.$$

Proof The theorem follows by defining

$$S_{m,n} = \sum_{j=0}^{m}\sum_{k=0}^{n} z_{j,k}$$

and

$$S_{m,n}^{*} = \sum_{j=0}^{m}\sum_{k=0}^{n} |z_{j,k}|$$

for all non-negative integers m and n, then noting that

$$|S_{p,q} - S_{m,n}| = \left| \sum_{j=0}^{m}\sum_{k=n+1}^{q} z_{j,k} + \sum_{j=m+1}^{p}\sum_{k=0}^{q} z_{j,k} \right|$$

$$\leq \sum_{j=0}^{m}\sum_{k=n+1}^{q} |z_{j,k}| + \sum_{j=m+1}^{q}\sum_{k=0}^{q} |z_{j,k}|$$

$$= S_{p,q}^{*} - S_{m,n}^{*}$$

whenever $p \geq m$ and $q \geq n$, and finally applying Theorem 1.6.2.   □

The proof of the next theorem is left as an exercise.

Theorem 1.6.5 *A double series $\sum_{j,k=0}^{\infty} a_{j,k}$, where $a_{j,k} \geq 0$ for all non-negative integers j and k, is convergent if and only if the set*

$$\left\{ \sum_{j=0}^{m} \sum_{k=0}^{n} a_{j,k} : m, n \in \mathbb{N} \right\}$$

is bounded, and in that case the series converges to the supremum of the set.

Theorem 1.6.6 *Let $\{c_n\}$ be an arrangement of a double sequence $\{z_{m,n}\}$. Then $\sum_{j,k=0}^{\infty} z_{j,k}$ converges absolutely if and only if $\sum_{j=0}^{\infty} c_j$ does so, and in that case*

$$\sum_{j,k=0}^{\infty} z_{j,k} = \sum_{j=0}^{\infty} c_j.$$

Proof For all non-negative integers m and n let

$$S_{m,n}^* = \sum_{j=0}^{m} \sum_{k=0}^{n} |z_{j,k}|$$

and

$$T_n^* = \sum_{j=0}^{n} |c_j|.$$

Suppose first that $\{S_{m,n}^*\}$ converges. For each l we can choose m and n large enough so that $S_{m,n}^*$ contains all the terms of T_l^*. Thus

$$0 \leq T_l^* \leq S_{m,n}^* \leq \lim_{(m,n)\to\infty} S_{m,n}^*.$$

Hence $\{T_n^*\}$ is non-decreasing and bounded above. It is therefore convergent. Similarly if $\{T_n^*\}$ is convergent then so is $\{S_{m,n}^*\}$.

Now suppose that $\{T_n^*\}$ converges to T^*. For all n define $T_n = \sum_{j=0}^{n} c_j$. Then the sequence $\{T_n\}$ converges to some number T. We must show that the double sequence $\{S_{m,n}\}$ also converges to T.

Choose $\varepsilon > 0$. There exist N such that $|T_N - T| < \varepsilon$ and $|T_N^* - T^*| < \varepsilon$. Choose m and n so large that $S_{m,n}$ contains all the terms of T_N. Let A be the set of terms in T_N, let B be the set of terms in $S_{m,n}$ but not in T_N and let C be the set of terms of $\{z_{j,k}\}$ that are not in $S_{m,n}$. Note that sets A and B are finite. As $\{T_n^*\}$ converges, so does $\{\sum_{x\in C} |x|\}$ and therefore so does $\{\sum_{x\in C} x\}$. Thus

$$T_N = \sum_{x \in A} x,$$

$$T_N^* = \sum_{x \in A} |x|,$$

$$T = \sum_{x \in A} x + \sum_{x \in B} x + \sum_{x \in C} x$$

and

$$T^* = \sum_{x \in A} |x| + \sum_{x \in B} |x| + \sum_{x \in C} |x|.$$

Moreover

$$S_{m,n} = T_N + \sum_{x \in B} x.$$

Hence

$$|S_{m,n} - T| \le |T_N - T| + \sum_{x \in B} |x|$$

$$< \varepsilon + \sum_{x \in B} |x| + \sum_{x \in C} |x|$$

$$= \varepsilon + T^* - T_N^*$$

$$< 2\varepsilon,$$

so that

$$\lim_{(m,n) \to \infty} S_{m,n} = T,$$

as required.   □

Example 1.6.4 Consider the double series

$$\sum_{j,k=1}^{\infty} \frac{1}{(j+k)^p},$$

where p is a positive number. Here we add the terms diagonal by diagonal. For each diagonal, $j + k$ is constant, and the diagonal for which $j + k = l$ has length $l - 1$. Hence

$$\sum_{j,k=1}^{\infty} \frac{1}{(j+k)^p} = \sum_{l=2}^{\infty} \frac{l-1}{l^p}.$$

For each $l \geq 2$ let

$$a_l = \frac{l-1}{l^p}$$

and

$$b_l = \frac{1}{l^{p-1}}.$$

Then

$$\frac{a_l}{b_l} = \frac{l-1}{l^p} \cdot l^{p-1} = \frac{l-1}{l} \to 1$$

as $l \to \infty$. By the limit comparison test and Theorem 1.6.6 we therefore find that the double series converges if and only if $p > 2$. $\triangle$

We turn next to iterated series.

Theorem 1.6.7 *If the double series $\sum_{j,k=0}^{\infty} z_{j,k}$ is absolutely convergent, then so are*

$$\sum_{j=0}^{\infty} \sum_{k=0}^{\infty} z_{j,k}$$

and

$$\sum_{k=0}^{\infty} \sum_{j=0}^{\infty} z_{j,k},$$

and the three series converge to the same limit.

Proof For all non-negative integers m and n, define

$$S_{m,n}^* = \sum_{j=0}^{m} \sum_{k=0}^{n} |z_{j,k}|$$

and

$$S_{m,n} = \sum_{j=0}^{m} \sum_{k=0}^{n} z_{j,k}.$$

Since $\sum_{j,k=0}^{\infty} z_{j,k}$ is absolutely convergent, the sequence $\{S_{m,n}^*\}$ converges to some number

$$S^* = \lim_{(m,n)\to\infty} S_{m,n}^*$$

and the sequence $\{S_{m,n}\}$ to some number

$$S = \lim_{(m,n)\to\infty} S_{m,n}.$$

Therefore for every $\varepsilon > 0$ there exists N such that

$$S^* - \varepsilon < S_{m,n}^* < S^* + \varepsilon$$

and

$$|S_{m,n} - S| < \varepsilon$$

whenever $m \geq N$ and $n \geq N$. For each fixed m it follows that the non-decreasing sequence $\{S_{m,n}^*\}$ is bounded above; let A_m be its limit. Thus

$$S^* - \varepsilon \leq A_m \leq S^* + \varepsilon.$$

Hence

$$\lim_{m\to\infty} A_m = S^*.$$

We have now proved that

$$\sum_{j=0}^{\infty}\sum_{k=0}^{\infty} |z_{j,k}| = \lim_{m\to\infty}\lim_{n\to\infty} \sum_{j=0}^{m}\sum_{k=0}^{n} |z_{j,k}| = S^*.$$

Similarly

$$\sum_{k=0}^{\infty}\sum_{j=0}^{\infty} |z_{j,k}| = S^*.$$

For each j let

$$B_j = \sum_{k=0}^{\infty} z_{j,k}.$$

For all $m \geq N$ and $n \geq N$ we have

$$\left| \sum_{j=0}^{m} B_j - S \right| = \left| \sum_{j=0}^{m} \sum_{k=0}^{\infty} z_{j,k} - S \right|$$

$$= \left| S_{m,n} + \sum_{j=0}^{m} \sum_{k=n+1}^{\infty} z_{j,k} - S \right|$$

$$\leq |S_{m,n} - S| + \sum_{j=0}^{m} \sum_{k=n+1}^{\infty} |z_{j,k}|$$

$$< \varepsilon + \sum_{j=0}^{m} \sum_{k=n+1}^{\infty} |z_{j,k}| + \sum_{j=m+1}^{\infty} \sum_{k=0}^{\infty} |z_{j,k}|$$

$$= \varepsilon + S^* - S^*_{m,n}$$

$$< 2\varepsilon.$$

Hence

$$\sum_{j=0}^{\infty} B_j = S.$$

We have now proved that

$$\sum_{j=0}^{\infty} \sum_{k=0}^{\infty} z_{j,k} = S.$$

The proof that

$$\sum_{k=0}^{\infty} \sum_{j=0}^{\infty} z_{j,k} = S$$

is similar.   □

Theorem 1.6.8 *If any of the series*

$$\sum_{j,k=0}^{\infty} z_{j,k}, \quad \sum_{j=0}^{\infty} \sum_{k=0}^{\infty} z_{j,k}, \quad \sum_{k=0}^{\infty} \sum_{j=0}^{\infty} z_{j,k}$$

converges absolutely, then all three series converge to the same limit.

Proof In view of Theorem 1.6.7, it suffices to show that if the second series is absolutely convergent then so is the first.

Let

$$\sum_{j=0}^{\infty}\sum_{k=0}^{\infty}|z_{j,k}| = S^*.$$

Let $\{c_n\}$ be an arrangement of the double sequence $\{|z_{m,n}|\}$. For each l there exist m and n so large that

$$\sum_{j=0}^{l}c_j \le \sum_{j=0}^{m}\sum_{k=0}^{n}|z_{j,k}| \le S^*.$$

Hence $\sum_{j=0}^{\infty}c_j$ is convergent. By Theorem 1.6.6, the double series $\sum_{j,k=0}^{\infty}|z_{j,k}|$ also converges. $\square$

Example 1.6.5 If $|x| < 1$ and $|y| < 1$, then Theorem 1.6.8 shows that

$$\sum_{j,k=0}^{\infty}|x^j y^k| = \sum_{j=0}^{\infty}\sum_{k=0}^{\infty}|x|^j|y|^k$$

$$= \sum_{j=0}^{\infty}\left(|x|^j \sum_{k=0}^{\infty}|y|^k\right)$$

$$= \sum_{j=0}^{\infty}\frac{|x|^j}{1-|y|}$$

$$= \frac{1}{(1-|x|)(1-|y|)}.$$

$\triangle$

Suppose that $\sum_{j=0}^{\infty}w_j$ and $\sum_{k=0}^{\infty}z_k$ converge absolutely. Since

$$\lim_{m\to\infty}\lim_{n\to\infty}\sum_{j=0}^{m}\sum_{k=0}^{n}|w_j z_k| = \lim_{m\to\infty}\lim_{n\to\infty}\sum_{j=0}^{m}|w_j|\sum_{k=0}^{n}|z_k|$$

$$= \sum_{j=0}^{\infty}|w_j|\sum_{k=0}^{\infty}|z_k|,$$

the iterated series

$$\sum_{j=0}^{\infty}\sum_{k=0}^{\infty}w_j z_k = \sum_{j=0}^{\infty}w_j\sum_{k=0}^{\infty}z_k$$

is absolutely convergent. Hence Theorem 1.6.8 shows that the double series

$$\sum_{j,k=0}^{\infty} w_j z_k$$

is absolutely convergent and converges to the same limit as the iterated series. By Theorem 1.6.6 the series

$$\sum_{j=0}^{\infty}\sum_{k=0}^{j} w_k z_{j-k}$$

also converges to this limit. We therefore deduce the following result.

Theorem 1.6.9 *If $\sum_{j=0}^{\infty} w_j$ and $\sum_{k=0}^{\infty} z_k$ are absolutely convergent series, then*

$$\sum_{j=0}^{\infty} w_j \sum_{k=0}^{\infty} z_k = \sum_{j=0}^{\infty}\sum_{k=0}^{j} w_k z_{j-k}, \qquad (1.6.5)$$

and the product is absolutely convergent.

The product given by Eq. (1.6.5) is called the **Cauchy product**.

Exercises 1.1

1. Let

$$a_{j,k} = \frac{(-1)^{j+k}}{2^{j+k-2}}\binom{j-1}{k-1}$$

for all positive integers m and n. Show that

$$\sum_{j=1}^{\infty} a_{j,k} = \sum_{k=1}^{\infty} a_{j,k} = \lim_{n\to\infty} S_{n,n} = 1$$

but the sequence $\{\Delta_n\}$ diverges.

2. For all non-negative integers m and n, let

$$a_{m,n} = \begin{cases} \frac{1}{m^2-n^2} & \text{if } m \neq n \\ 0 & \text{otherwise.} \end{cases}$$

Show that

$$\sum_{k=0}^{\infty}\sum_{j=0}^{\infty} a_{j,k} = -\sum_{j=0}^{\infty}\sum_{k=0}^{\infty} a_{j,k}.$$

3. For the double sequence $\{a_{m,n}\}$ given by the array

$$
\begin{array}{ccccc}
1 & 1 & 1 & 1 & \cdots \\
1 & -1 & -1 & -1 & \cdots \\
1 & -1 & 0 & 0 & \cdots \\
1 & -1 & 0 & 0 & \cdots \\
\vdots & \vdots & \vdots & \vdots & \ddots
\end{array},
$$

discuss the convergence of each of the following series:

(a) $\sum_{j,k=0}^{\infty} a_{j,k}$;
(b) $\sum_{j=0}^{\infty} \sum_{k=0}^{\infty} a_{j,k}$;
(c) $\sum_{k=0}^{\infty} \sum_{j=0}^{\infty} a_{j,k}$;
(d) $\sum_{k=0}^{\infty} \Delta_k$.

4. For the double sequence $\{a_{m,n}\}$ given by the array

$$
\begin{array}{ccccc}
0 & 1 & 0 & 0 & \cdots \\
-1 & 0 & 1 & 0 & \cdots \\
0 & -1 & 0 & 1 & \cdots \\
0 & 0 & -1 & 0 & \cdots \\
\vdots & \vdots & \vdots & \vdots & \ddots
\end{array},
$$

calculate:

(a) $S_{m,n}$ for all non-negative integers m and n;
(b) $\sum_{j=0}^{\infty} \sum_{k=0}^{\infty} a_{j,k}$;
(c) $\sum_{k=0}^{\infty} \sum_{j=0}^{\infty} a_{j,k}$.

5. Evaluate:

(a) $\sum_{j,k=1}^{\infty} \frac{1}{j^k}$;
(b) $\sum_{j,k=1}^{\infty} \frac{1}{(p+j)^k}$, where $p > -1$;
(c) $\sum_{j,k=1}^{\infty} \frac{1}{(2j)^k}$.

6. Test the convergence of the following double series, where $\alpha > 1$ and $\beta > 1$:

(a) $\sum_{j,k=1}^{\infty} \frac{1}{j^\alpha k^\beta}$;
(b) $\sum_{j,k=1}^{\infty} \frac{1}{(j^2+k^2)^\alpha}$.

7. Show that

$$
\sum_{j,k=1}^{\infty} \frac{1}{j^2 \left(1 + k^{\frac{3}{2}}\right)}
$$

converges.

8. Let $a = \sum_{j=0}^{\infty} a_j$ and $b = \sum_{j=0}^{\infty} b_j$ be absolutely convergent complex series and define $z_{m,n} = a_m b_n$ for each m and n. Show that $\sum_{m,n=0}^{\infty} z_{m,n}$ is absolutely convergent to ab.

Chapter 2
Infinite Products

The theory of sequences can be combined with the familiar notion of a finite product to produce a theory of infinite products. The theory of infinite series shadows most of this material because many of the convergence questions for infinite products can be answered by results from the theory of series. The theory of infinite products nonetheless has a flavour distinct from that of series, as becomes evident once the definitions of absolute and conditional convergence of products are introduced.

The general theory for the convergence of infinite products of numbers is presented in the first three sections of this chapter.

The representation of functions as infinite products looms large in analysis. Sects. 2.4 and 2.5 concern infinite products of functions. A highlight is a product representation for $\sin x$ with factors $\cos(x/2^n)$ derived in Example 2.5.2. This representation exploits the double angle formula for $\sin x$ and leads to Viète's formula for π (Eq. (2.5.7)). In Sect. 2.6 product representations are derived for $\sin x$ and $\cos x$ using elementary techniques from calculus. The representation for $\sin x$ yields the Wallis product for π (Eq. (2.6.8)) and a formula for the evaluation of the Riemann zeta function at even integers (Eq. (2.6.17)). An analogue of Abel's theorem is given in Sect. 2.7 for infinite products. *Prima facie* it seems the extension of this theorem to infinite products is straightforward, yet it is here that the distinction between regularly and irregularly convergent products is felt. A feature of this section is an example due to Hardy (cf. Eq. (2.7.6)) that is counterintuitive.

The product representations of $\sin x$ and $\cos x$ given in Sect. 2.6 directly identify the zeros of these functions through the factors of the product. This observation motivates the question of whether a function with a given set of zeros can be represented by an infinite product with factors that identify the zeros. The Weierstrass factorization theorem (Theorem 2.9.2) answers this question for entire functions. Theorem 2.9.3 answers it for functions that are analytic in the unit disc $D(0; 1)$. The chapter ends with a discussion of double infinite products.

© The Author(s), under exclusive license to Springer Nature Switzerland AG 2022   39
C. H. C. Little et al., *An Introduction to Infinite Products*, SUMS Readings,
https://doi.org/10.1007/978-3-030-90646-7_2

2.1 Introduction

Let $\{z_n\}$ be a sequence of complex numbers, and let

$$\prod_{j=1}^{n} z_j = z_1 z_2 \cdots z_n$$

for each $n \in \mathbb{N}$. The **sequence of partial products** $\{P_n\}$ is defined by

$$P_n = \prod_{j=1}^{n} z_j,$$

for each $n \in \mathbb{N}$. The limiting case, as $n \to \infty$, is called an **infinite product** and denoted

$$\prod_{j=1}^{\infty} z_j.$$

The numbers $z_1, z_2, \ldots$ are called the **factors** of the product.

The definitions of convergence and divergence for infinite products are analogous to those for infinite series with the sequence of partial products replacing the sequence of partial sums. Specifically, if the sequence $\{P_n\}$ diverges, then the infinite product is said to **diverge**. If $\{P_n\}$ converges to a *nonzero* limit, then the infinite product is said to **converge**. The last definition seems peculiar at first glance because the case $P_n \to 0$ as $n \to \infty$ has been excluded. If we denote by $\log z$ the logarithm of a complex number z, then we will see shortly that the convergence or divergence of the product is determined by the convergence or divergence of the series $\sum_{j=1}^{\infty} \log z_j$. The case $P_n \to 0$ corresponds to divergence to $-\infty$ for the series. If $P_n \to 0$, there are two cases that we distinguish. It is clear that if z_k is zero for some $k \in \mathbb{N}$ then $P_n \to 0$. If there are a *finite* but positive number of factors that are zero, these factors may be removed and we can study the remaining product. If this product is convergent as defined above then the original product is said to **converge to zero**. If $P_n \to 0$ and the product does not converge to zero, then it is said to **diverge to zero**.

Example 2.1.1 Consider the product

$$\prod_{j=1}^{\infty} \frac{1}{2^{1/2^j}}.$$

The partial products are given by

$$P_n = \frac{1}{2^{1/2}} \cdot \frac{1}{2^{1/4}} \cdots \frac{1}{2^{1/2^n}}$$
$$= 2^{-\sum_{j=1}^{n}(1/2)^j}$$
$$= 2^{-1+1/2^n}.$$

Evidently $P_n \to 1/2$ as $n \to \infty$ and therefore the product converges.   △

Example 2.1.2 Consider the product

$$\prod_{j=1}^{\infty} \frac{1}{2^j}.$$

Here,

$$P_n = \frac{1}{2} \cdot \frac{1}{2^2} \cdots \frac{1}{2^n}$$
$$= 2^{-(1+2+\cdots+n)}$$
$$= 2^{-\frac{n(n+1)}{2}};$$

consequently, $P_n \to 0$ as $n \to \infty$. The product thus diverges to zero.   △

If $\{z_n\}$ and $\{w_n\}$ are sequences of numbers such that $\prod_{j=1}^{\infty} z_j$ converges to $A \neq 0$ and $\prod_{j=1}^{\infty} w_j$ to $B \neq 0$, then $\prod_{j=1}^{\infty} z_j w_j$ converges to

$$\lim_{n\to\infty} \prod_{j=1}^{n} z_j w_j = \lim_{n\to\infty} \prod_{j=1}^{n} z_j \prod_{j=1}^{n} w_j = \prod_{j=1}^{\infty} z_j \prod_{j=1}^{\infty} w_j = AB \neq 0.$$

Similarly if $w_n \neq 0$ for all n, then $\prod_{j=1}^{\infty}(z_j/w_j)$ converges to $A/B \neq 0$.

A convergent product can contain a finite number of negative factors, but it is clear that it cannot have infinitely many negative factors. Indeed, if $\prod_{j=1}^{\infty} w_j$ converges, it is intuitively obvious that $w_n \to 1$ as $n \to \infty$. More formally suppose that $\prod_{j=1}^{\infty} w_j$ converges (to a nonzero number). Now,

$$P_n = w_n P_{n-1}$$

for all $n > 1$, and since $P_n \neq 0$ for all $n \in \mathbb{N}$,

$$w_n = \frac{P_n}{P_{n-1}};$$

therefore,

$$\lim_{n\to\infty} w_n = \lim_{n\to\infty} \frac{P_n}{P_{n-1}} = 1.$$

This result motivates the standard notation for infinite products,

$$\prod_{j=1}^{\infty}(1 + z_j).$$

The argument above provides a proof for the analogue of the nth term test for series.

Theorem 2.1.1 (nth Term Test) *Let $\{z_n\}$ be a sequence of numbers. If the product*

$$\prod_{j=1}^{\infty}(1 + z_j)$$

converges, then $z_n \to 0$ as $n \to \infty$.

Exercises 2.1

1. Show that

$$\sum_{j=1}^{\infty} \frac{j}{2^j} = 2,$$

   and therefore

$$\prod_{j=1}^{\infty} 2^{j/2^j} = 4.$$

   (Hint: Differentiate the geometric series.)
2. Show that

$$\prod_{j=2}^{n}\left(1 - \frac{1}{j^2}\right) = \frac{n+1}{2n},$$

   and hence

$$\prod_{j=2}^{\infty}\left(1 - \frac{1}{j^2}\right) = \frac{1}{2}.$$

   Similarly show that

$$\prod_{j=3}^{\infty}\left(1 - \frac{4}{j^2}\right) = \frac{1}{6}$$

and, in general, that

$$\prod_{j=k+1}^{\infty}\left(1 - \frac{k^2}{j^2}\right) = \frac{1}{\binom{2k}{k}}$$

for any $k > 0$.

3. Use the algebraic identities

$$j^3 - 1 = (j - 1)(j^2 + j + 1)$$

and

$$j^3 + 1 = (j + 1)(j^2 - j + 1)$$

to show that the partial products for

$$P = \prod_{j=2}^{\infty} \frac{j^3 - 1}{j^3 + 1}$$

satisfy

$$P_{n+1} = \frac{2(n^2 + 3n + 3)}{3(n + 1)(n + 2)},$$

and thereby deduce that $P = 2/3$.

4. Given the convergence of $\prod_{j=1}^{\infty} z_j$ and $\prod_{j=1}^{\infty} w_j$, investigate the convergence of each of the following products:

(a) $\prod_{j=1}^{\infty}(z_j + w_j)$,
(b) $\prod_{j=1}^{\infty} z_j w_j$,
(c) $\prod_{j=1}^{\infty} \frac{z_j}{w_j}$.

5. Show that

$$\prod_{j=1}^{n}\left(1 + z^{2^j}\right) = \frac{1 - z^{2^{n+1}}}{1 - z}$$

for each $n \in \mathbb{N}$ and $z \neq 1$ and hence investigate the convergence of the corresponding infinite product.

6. Show that

$$\prod_{j=1}^{\infty} \left(1 + \frac{1}{j(j+2)} \right) = 2.$$

7. Show that

$$\prod_{j=1}^{n} \frac{e^{\frac{1}{j}}}{1 + \frac{1}{j}} = \frac{n}{n+1} \exp \left(\sum_{j=1}^{n} \frac{1}{j} - \log n \right)$$

for all $n \in \mathbb{N}$ and hence find the value of

$$\prod_{j=1}^{\infty} \frac{e^{\frac{1}{j}}}{1 + \frac{1}{j}}.$$

8. For each $a > 0$, find the value of $\prod_{j=1}^{\infty} a^{(-1)^j/j}$.

2.2 Convergence of Products and Series

The convergence of a real product $\prod_{j=1}^{\infty}(1 + a_j)$ is closely linked with that of the series $\sum_{j=1}^{\infty} a_j$. This section is concerned with establishing that, if the a_j do not change sign, then the series and the product either both converge or both diverge. In addition, the definition of absolute convergence for a product is introduced and related to absolute convergence for a series.

Theorem 2.2.1 *Let $\{a_n\}$ be a sequence of non-negative numbers. Then the series $\sum_{j=1}^{\infty} a_j$ and the product $\prod_{j=1}^{\infty}(1 + a_j)$ either both converge or both diverge.*

Proof If $x \geq 0$, then $1 + x \leq e^x$; hence for each $n \in \mathbb{N}$ we have

$$a_1 + a_2 + \cdots + a_n < (1 + a_1)(1 + a_2) \cdots (1 + a_n)$$
$$\leq e^{a_1 + a_2 + \cdots + a_n},$$

so that

$$S_n < P_n \leq e^{S_n},$$

where $S_n = \sum_{j=1}^{n} a_j$ and $P_n = \prod_{j=1}^{n}(1 + a_j)$. Note that both $\{S_n\}$ and $\{P_n\}$ are non-decreasing sequences.

Suppose that the series converges to S. Then $S_n \to S$ as $n \to \infty$ and $S_n \leq S$ for all n. Therefore,

$$P_n \leq e^{S_n} \leq e^S,$$

and consequently $\{P_n\}$ is bounded above. Thus $\{P_n\}$ converges to some limit P. Since $0 < P_1 \leq P_n \leq P$, we have $P \neq 0$ and therefore the product converges.

Suppose now that $\{P_n\}$ converges to a limit P. Then $P_n \leq P$ for all $n \in \mathbb{N}$. The sequence $\{S_n\}$ is non-decreasing and bounded above by P; hence, $\{S_n\}$ converges and therefore the series converges.   □

Example 2.2.1   Theorem 2.2.1 provides a simple proof that the harmonic series is divergent. Note that

$$\prod_{j=1}^{n}\left(1 + \frac{1}{j}\right) = \prod_{j=1}^{n} \frac{j+1}{j}$$

$$= n + 1.$$

Thus the product

$$\prod_{j=1}^{\infty}\left(1 + \frac{1}{j}\right)$$

is divergent. It follows by Theorem 2.2.1 that the harmonic series is divergent.   △

Theorem 2.2.2   *Let $\{a_n\}$ be a sequence of numbers such that $-1 < a_n \leq 0$ for all $n \in \mathbb{N}$. Then the series $\sum_{j=1}^{\infty} a_j$ and the product $\prod_{j=1}^{\infty}(1+a_j)$ either both converge or both diverge.*

Proof   We use the inequality

$$1 - x \leq e^{-x}, \tag{2.2.1}$$

which holds for all real x. For all $n \in \mathbb{N}$ let $b_n = -a_n$, $S_n = \sum_{j=1}^{n} b_j$ and

$$P_n = \prod_{j=1}^{n}(1 - b_j).$$

Then $0 < 1 - b_j \leq 1$ for each j, and inequality (2.2.1) implies that

$$0 < P_n \leq e^{-S_n}. \tag{2.2.2}$$

Thus $\{P_n\}$ is a non-increasing sequence bounded below by 0. The sequence therefore converges to a non-negative limit. If the product diverges, it must diverge to 0.

We show first that if the series diverges then so does the product. Suppose that $\{S_n\}$ diverges. This sequence is non-decreasing and therefore $S_n \to \infty$ as $n \to \infty$.

Inequality (2.2.2) thus implies that $P_n \to 0$ as $n \to \infty$; hence the product diverges to zero as none of its factors is 0.

Suppose now that $\sum_{j=1}^{\infty} a_j$ converges. Then $\{S_n\}$ converges and therefore there is an integer N such that

$$\sum_{j=N+1}^{n} b_j = S_n - S_N < \frac{1}{2} \tag{2.2.3}$$

for all $n > N$, by the Cauchy principle.

Since $b_j < 1$ for all j it is clear that $P_N \neq 0$. We show that

$$\frac{P_n}{P_N} \geq 1 - \sum_{j=N+1}^{n} b_j \tag{2.2.4}$$

for all $n > N$. First, both sides are equal to $1 - b_{N+1}$ for $n = N + 1$. Suppose that inequality (2.2.4) is true for $n = N + k$. Then

$$\begin{aligned}
\frac{P_{N+k+1}}{P_N} &= \frac{P_{N+k}}{P_N}(1 - b_{N+k+1}) \\
&\geq \left(1 - \sum_{j=N+1}^{N+k} b_j\right)(1 - b_{N+k+1}) \\
&= 1 - \sum_{j=N+1}^{N+k+1} b_j + b_{N+k+1} \sum_{j=N+1}^{N+k} b_j \\
&\geq 1 - \sum_{j=N+1}^{N+k+1} b_j.
\end{aligned}$$

Inequality (2.2.4) therefore follows by induction. The sequence $\{P_n/P_N\}$ is non-increasing, and inequalities (2.2.3) and (2.2.4) show that

$$\frac{P_n}{P_N} > 1 - \frac{1}{2} = \frac{1}{2}.$$

The sequence $\{P_n/P_N\}$ therefore converges to a nonzero limit and hence $\{P_n\}$ must converge to a nonzero limit. We conclude that if the series converges then the product converges. $\qquad\square$

Corollary 2.2.3 *Let $\{a_j\}$ be a sequence of non-negative real numbers. Then the series $\sum_{j=1}^{\infty} a_j$ and the product $\prod_{j=1}^{\infty}(1 - a_j)$ either both converge or both diverge.*

Proof If either $\sum_{j=1}^{\infty} a_j$ or $\prod_{j=1}^{\infty}(1 - a_j)$ is convergent, then $\lim_{n \to \infty} a_n = 0$. Hence there exists $N > 0$ such that $a_n < 1$ for all $n \geq N$. Since $-1 < -a_n \leq 0$ for all such n, the result is immediate from Theorem 2.2.2. $\qquad \square$

Example 2.2.2 We shall show that $\prod_{j=1}^{\infty} j \sin(1/j)$ converges. First we rewrite the product as

$$\prod_{j=1}^{\infty} \left(1 - \left(1 - j \sin \frac{1}{j}\right)\right).$$

Since

$$\sin \frac{1}{j} \leq \frac{1}{j},$$

we have

$$j \sin \frac{1}{j} \leq 1,$$

so that

$$1 - j \sin \frac{1}{j} \geq 0.$$

Therefore Corollary 2.2.3 can be applied.

Using L'Hôpital's rule we compute

$$\lim_{x \to 0} \frac{1 - \frac{\sin x}{x}}{x^2} = \lim_{x \to 0} \frac{x - \sin x}{x^3}$$
$$= \lim_{x \to 0} \frac{1 - \cos x}{3x^2}$$
$$= \lim_{x \to 0} \frac{\sin x}{6x}$$
$$= \lim_{x \to 0} \frac{\cos x}{6}$$
$$= \frac{1}{6}.$$

By setting $x = 1/j$, we therefore find that

$$\lim_{j \to \infty} \frac{1 - j \sin \frac{1}{j}}{\frac{1}{j^2}} = \frac{1}{6}.$$

Thus

$$1 - j \sin \frac{1}{j} \sim \frac{1}{6j^2}.$$

Since $\sum_{j=1}^{\infty}(1/j^2)$ converges, so does

$$\sum_{j=1}^{\infty}\left(1 - j \sin \frac{1}{j}\right),$$

by the limit comparison test. The convergence of the given product now follows from Corollary 2.2.3.   △

Example 2.2.3 The product

$$\prod_{j=1}^{\infty}\left(1 + \frac{1}{j^2}\right)$$

converges by Theorem 2.2.1, since $a_j = 1/j^2$ and the series $\sum_{j=1}^{\infty} a_j$ is a convergent p-series, with $p = 2$. In contrast, the product

$$\prod_{j=2}^{\infty}\left(1 - \frac{1}{j}\right)$$

diverges by Corollary 2.2.3 because the series $\sum_{j=2}^{\infty} 1/j$ diverges. In fact, we can deduce this result directly since

$$\begin{aligned}
P_n &= \left(1 - \frac{1}{2}\right)\left(1 - \frac{1}{3}\right) \cdots \left(1 - \frac{1}{n}\right) \\
&= \left(\frac{1}{2}\right)\left(\frac{2}{3}\right) \cdots \left(\frac{n-2}{n-1}\right)\left(\frac{n-1}{n}\right) \\
&= \frac{1}{n},
\end{aligned}$$

so that $P_n \to 0$ as $n \to \infty$.   △

For the purposes of determining whether a product $\prod_{j=1}^{\infty}(1 + a_j)$ converges or diverges we can always ignore any finite number of nonzero factors. We can thus apply Theorem 2.2.1 if there is an $N \in \mathbb{N}$ such that $a_j \neq -1$ whenever $1 \leq j \leq N$ and $a_j \geq 0$ for all $j > N$. A similar comment holds for Theorem 2.2.2.

Example 2.2.4 Consider the product

$$P(\theta) = \prod_{k=1}^{\infty} \left(1 - k \sin \frac{\theta}{k^2}\right).$$

Evidently, this product converges for $\theta = 0$. Suppose that $\theta \neq 0$. The Maclaurin series for $\sin(\theta/k^2)$ gives

$$k \sin \frac{\theta}{k^2} = \sum_{j=0}^{\infty} \frac{(-1)^j \theta^{2j+1}}{(2j+1)! k^{4j+1}} \tag{2.2.5}$$

for all $k \in \mathbb{N}$. For some values of θ there is a k such that $k \sin(\theta/k^2) = 1$. For instance, if $\theta = \pi/2$ then $k \sin(\theta/k^2) = 1$ when $k = 1$. However only a finite number of factors can vanish, for Eq. (2.2.5) shows that $k \sin(\theta/k^2) \to 0$ as $k \to \infty$. Moreover $\sin(\theta/k^2)$ has the sign of θ if $|\theta| < \pi k^2$. Since $\pi k^2 \to \infty$ as $k \to \infty$, there is an $N \in \mathbb{N}$ such that if $k > N$ then the sign of $k \sin(\theta/k^2)$ is that of θ and $|k \sin(\theta/k^2)| < 1$. We can thus ignore the first N factors where sign changes (if any) occur and use Theorem 2.2.1 if $\theta < 0$ or Theorem 2.2.2 if $\theta > 0$.

Equation (2.2.5) gives

$$\begin{aligned}
k \sin \frac{\theta}{k^2} &= \frac{\theta}{k} \sum_{j=0}^{\infty} \frac{(-1)^j \theta^{2j}}{(2j+1)! k^{4j}} \\
&= \frac{\theta}{k} \left(1 + \sum_{j=1}^{\infty} \frac{(-1)^j \theta^{2j}}{(2j+1)! k^{4j}}\right) \\
&\sim \frac{\theta}{k}.
\end{aligned}$$

The limit comparison test (Theorem 1.2.2) therefore shows that

$$\sum_{k=1}^{\infty} k \sin \frac{\theta}{k^2}$$

diverges. If $\theta < 0$, Theorem 2.2.1 implies that the product diverges; if $\theta > 0$, Theorem 2.2.2 (or Corollary 2.2.3) implies that the product diverges. The product thus diverges for all $\theta \neq 0$. △

How should one define the absolute convergence of an infinite product? A naïve approach would be to say that the product $\prod_{j=1}^{\infty} z_j$ is absolutely convergent if $\prod_{j=1}^{\infty} |z_j|$ is convergent. One might then expect that absolute convergence implies convergence, as for infinite series. The following example shows that this is not the case.

Example 2.2.5 We show that the product

$$\prod_{j=1}^{\infty} \left| 1 + \frac{i}{j} \right|$$

is convergent whereas

$$\prod_{j=1}^{\infty} \left(1 + \frac{i}{j} \right)$$

is divergent.

Example 2.2.3 shows that

$$\prod_{j=1}^{\infty} \left(1 + \frac{1}{j^2} \right)$$

converges. Since

$$\prod_{j=1}^{\infty} \left| 1 + \frac{i}{j} \right| = \prod_{j=1}^{\infty} \left(1 + \frac{1}{j^2} \right)^{1/2}$$

$$= \left(\prod_{j=1}^{\infty} \left(1 + \frac{1}{j^2} \right) \right)^{1/2},$$

so does the first of the given products.

On the other hand,

$$1 + \frac{i}{j} = \left| 1 + \frac{i}{j} \right| e^{i\theta_j}$$

for all $j > 0$, where

$$\theta_j = \arctan \frac{1}{j}.$$

Hence

$$\prod_{j=1}^{\infty} \left(1 + \frac{i}{j} \right) = \left(\prod_{j=1}^{\infty} \left| 1 + \frac{i}{j} \right| \right) \left(\prod_{j=1}^{\infty} e^{i\theta_j} \right)$$

$$= \left(\prod_{j=1}^{\infty} \left| 1 + \frac{i}{j} \right| \right) \exp \left(i \sum_{j=1}^{\infty} \theta_j \right).$$

We have already shown that the product on the right hand side converges. In order to establish the divergence of the product on the left hand side, it is therefore enough to demonstrate that $\sum_{j=1}^{\infty} \theta_j$ diverges.

From the Maclaurin series for the inverse tangent function and Theorem 1.3.8 we obtain

$$\left| \theta_j - \frac{1}{j} \right| \le \frac{1}{3 j^3}.$$

Since $\sum_{j=1}^{\infty} 1/(3 j^3)$ converges, so does

$$\sum_{j=1}^{\infty} \left| \theta_j - \frac{1}{j} \right|$$

by the comparison test. Therefore

$$\sum_{j=1}^{\infty} \left(\theta_j - \frac{1}{j} \right)$$

converges. However the harmonic series diverges. We conclude that $\sum_{j=1}^{\infty} \theta_j$ diverges, as required. △

A product $\prod_{j=1}^{\infty} (1 + z_j)$ is called **absolutely convergent** if the product

$$\prod_{j=1}^{\infty} (1 + |z_j|) \tag{2.2.6}$$

is convergent. The product is called **conditionally convergent** if $\prod_{j=1}^{\infty} (1 + z_j)$ converges but the product (2.2.6) diverges.

Our next goal is to investigate the convergence of an absolutely convergent product. Recall that if $\prod_{j=1}^{\infty} (1 + z_j)$ converges, then $z_j \to 0$ as $j \to \infty$. Thus there exists N such that $z_j \ne -1$ for all $j \ge N$. This is a necessary condition for convergence. We shall show that if it holds and $\prod_{j=1}^{\infty} (1 + |z_j|)$ converges, then so does $\prod_{j=1}^{\infty} (1 + z_j)$. It is enough to consider the case where $z_j \ne -1$ for all j.

The proof that an absolutely convergent product $\prod_{j=1}^{\infty} (1 + z_j)$ is convergent if $z_j \ne -1$ for all j depends on the logarithmic series associated with the product. Let

$$P_n = \prod_{j=1}^{n}(1 + z_j),$$

and consider the sequence defined by

$$\log P_n = \sum_{j=1}^{n} \log\left(1 + z_j\right).$$

If the product converges, then the limit P is nonzero and the sequence $\{\log P_n\}$ converges to $\log P$. Conversely, if $\{\log P_n\}$ converges to L, then the product converges to $P = e^L > 0$, since $P_n = e^{\log P_n}$ for all n. In fact, if we define $\log z$ to be the principal branch of the logarithm of z for any complex $z \neq 0$, so that

$$\log z = \log |z| + i\theta$$

if $z = |z|e^{i\theta}$ where $\theta \in (-\pi, \pi]$, then we have the following theorem.

Theorem 2.2.4 *If $z_j \neq 0$ for all j, then the product $\prod_{j=1}^{\infty} z_j$ converges if and only if $\sum_{j=1}^{\infty} \log z_j$ does so. Moreover, if the series converges to S, then the product converges to e^S.*

Proof For all n let $P_n = \prod_{j=1}^{n} z_j$ and

$$S_n = \sum_{j=1}^{n} \log z_j = \log P_n + 2\pi k_n i$$

for some integer k_n. Then

$$e^{S_n} = e^{\log P_n} e^{2\pi k_n i} = P_n \cdot 1 = P_n.$$

It follows that if the series converges to S then the product converges to $e^S \neq 0$.

Suppose on the other hand that the product converges to $P \neq 0$. For all $n > 1$ we have

$$\log \frac{P_n}{P_{n-1}} = \log z_n$$

$$= S_n - S_{n-1}$$

$$= \log \frac{P_n}{P_{n-1}} + 2\pi (k_n - k_{n-1})i.$$

Therefore $k_n = k_{n-1}$, and so there exists a constant k such that

$$S_n = \log P_n + 2k\pi i.$$

Since $P \neq 0$, the series converges to $\log P + 2k\pi i$. □

Thus if $\prod_{j=1}^{\infty} z_j = P$, then

$$\sum_{j=1}^{\infty} \log z_j = \log P + 2k\pi i \qquad (2.2.7)$$

for some integer k. Moreover, if $\sum_{j=1}^{\infty} \log z_j = S$ then $\prod_{j=1}^{\infty} z_j = e^S$.

The next example illustrates the possibility that $k \neq 0$ in Eq. (2.2.7).

Example 2.2.6 For the product

$$\prod_{j=1}^{\infty} \exp\left(\frac{-\pi i}{2^j}\right)$$

and any positive integer n we have

$$
\begin{aligned}
P_n &= \prod_{j=1}^{n} \exp\left(\frac{-\pi i}{2^j}\right) \\
&= \exp\left(-\pi i \sum_{j=1}^{n} \frac{1}{2^j}\right) \\
&= \exp\left(-\pi i \left(1 - \frac{1}{2^n}\right)\right) \\
&\to e^{-\pi i} \\
&= e^{\pi i}
\end{aligned}
$$

as $n \to \infty$. Letting $P = e^{\pi i}$, we obtain $\log P = \pi i$.

On the other hand, since $-\pi/2^j \in (-\pi, \pi]$ we find that

$$
\begin{aligned}
\sum_{j=1}^{n} \log \exp\left(\frac{-\pi i}{2^j}\right) &= -\pi i \sum_{j=1}^{n} \frac{1}{2^j} \\
&= -\pi i \left(1 - \frac{1}{2^n}\right).
\end{aligned}
$$

Therefore

$$\sum_{j=1}^{\infty} \log \exp\left(\frac{-\pi i}{2^j}\right) = -\pi i$$

$$= \log P - 2\pi i.$$

$$\triangle$$

Our investigation of the convergence of an absolutely convergent product requires the next result, which is of independent interest.

Theorem 2.2.5 *Let* $z_j \neq -1$ *for all* $j \in \mathbb{N}$. *If the series* $\sum_{j=1}^{\infty} |z_j|$ *converges, then the series* $\sum_{j=1}^{\infty} |\log(1 + z_j)|$ *converges.*

Proof For $|z| < 1$ we have

$$|\log(1+z)| = \left| \sum_{j=0}^{\infty} (-1)^j \frac{z^{j+1}}{j+1} \right|$$

$$\leq |z| \sum_{j=0}^{\infty} |z|^j$$

$$= \frac{|z|}{1 - |z|}. \tag{2.2.8}$$

Now, $\sum_{j=0}^{\infty} |z_j|$ is convergent and so there is an integer N such that $|z_j| < 1/2$ for all $j \geq N$. For all such j we consequently have

$$|\log(1 + z_j)| \leq \frac{|z_j|}{1 - |z_j|} < 2|z_j|.$$

The result therefore follows from the comparison test. □

Theorem 2.2.6 *Let* $z_j \neq -1$ *for all* $j \in \mathbb{N}$. *If the product* $\prod_{j=1}^{\infty}(1+|z_j|)$ *converges, then so does* $\prod_{j=1}^{\infty}(1 + z_j)$.

Proof Since $\prod_{j=1}^{\infty}(1 + |z_j|)$ converges, Theorem 2.2.1 implies that $\sum_{j=1}^{\infty} |z_j|$ is convergent, and Theorem 2.2.5 thus shows that $\sum_{j=1}^{\infty} |\log(1 + z_j)|$ converges. The series $\sum_{j=1}^{\infty} \log(1+z_j)$ therefore converges and consequently so does $\prod_{j=1}^{\infty}(1+z_j)$ by Theorem 2.2.4. □

The following test is useful when the terms of a real sequence change sign.

Corollary 2.2.7 *Let* $\{z_j\}$ *be a sequence such that* $z_j \neq -1$ *for all* $j \in \mathbb{N}$ *and the series* $\sum_{j=1}^{\infty} |z_j|$ *converges. Then the product* $\prod_{j=1}^{\infty}(1 + z_j)$ *is convergent (to a nonzero number).*

Proof As $\sum_{j=1}^{\infty} |z_j|$ converges, the product is absolutely convergent by Theorem 2.2.1. The result now follows by Theorem 2.2.6. □

In the next section we give an example (Example 2.3.1) where the converse of this corollary does not hold.

Example 2.2.7 Let α and ϕ be real numbers such that $\alpha > 1$. Define functions C and S by

$$C(\phi, \alpha) = \prod_{j=1}^{\infty} \left(1 + \frac{\cos j\phi}{j^{\alpha}}\right)$$

and

$$S(\phi, \alpha) = \prod_{j=1}^{\infty} \left(1 + \frac{\sin j\phi}{j^{\alpha}}\right).$$

We show that these products converge to a nonzero number for all $\phi \in \mathbb{R}$ except for some integer multiples of $\pi/2$ where they converge to 0.

Let

$$c_j = \frac{\cos j\phi}{j^{\alpha}}$$

and

$$s_j = \frac{\sin j\phi}{j^{\alpha}}.$$

Then, for all $j \in \mathbb{N}$ and $\phi \in \mathbb{R}$, it follows that $|c_j| \leq 1/j^{\alpha}$ and $|s_j| \leq 1/j^{\alpha}$. Since $\alpha > 1$, the series $\sum_{j=1}^{\infty} 1/j^{\alpha}$ converges and therefore the series $\sum_{j=1}^{\infty} c_j$ and $\sum_{j=1}^{\infty} s_j$ converge absolutely.

Corollary 2.2.7 shows that the product C converges to a nonzero number provided $c_j \neq -1$ for all j. Suppose that $c_j = -1$ for some j. Then $\cos j\phi = -j^{\alpha}$. Since $\cos j\phi \geq -1$ we must have $j = 1$ and therefore $\cos \phi = -1$. We thus see that only the first factor can vanish and for this case $\phi = (2m + 1)\pi$, where m is an integer. For this set of values for ϕ the product converges to zero since the previous arguments show that

$$\prod_{j=2}^{\infty} \left(1 + \frac{\cos j\phi}{j^{\alpha}}\right)$$

must be convergent for all $\phi \in \mathbb{R}$. The result for the product S follows in a similar manner. $\triangle$

Theorem 2.2.4 may be used to give an alternative proof of Theorem 2.2.1.

Theorem 2.2.8 *Series $\sum_{j=1}^{\infty} z_j$ converges absolutely if and only if $\prod_{j=1}^{\infty}(1 + z_j)$ converges absolutely.*

Proof We shall show that $\sum_{j=1}^{\infty} |z_j|$ and $\prod_{j=1}^{\infty}(1 + |z_j|)$ both converge or both diverge. According to Theorem 2.2.4 it is enough to show that the series $\sum_{j=1}^{\infty} |z_j|$ and $\sum_{j=1}^{\infty} \log(1+|z_j|)$ both converge or both diverge. Both diverge unless $|z_j| \to 0$ as $j \to \infty$. We may therefore assume that $0 < |z_j| < 1$ for each j. The Maclaurin series for $\log(1 + |z_j|)$ consequently shows that

$$\frac{\log(1 + |z_j|)}{|z_j|} \to 1$$

as $j \to \infty$. The desired result now follows from the limit comparison test. $\qquad\square$

Theorem 2.2.4 may also be used to establish the divergence of the product

$$\prod_{j=1}^{\infty} \left(1 + \frac{i}{j}\right)$$

studied in Example 2.2.5. For all $j \geq 2$ the Maclaurin series for $\log(1 + i/j)$ gives

$$\left|\log\left(1 + \frac{i}{j}\right) - \frac{i}{j}\right| = \left|\sum_{k=1}^{\infty}(-1)^k \frac{i^{k+1}}{(k + 1) j^{k+1}}\right|$$

$$\leq \frac{1}{j^2} \sum_{k=1}^{\infty} \frac{1}{(k + 1) j^{k-1}}$$

$$< \frac{1}{j^2} \sum_{k=0}^{\infty} \frac{1}{j^k}$$

$$= \frac{1}{j^2} \cdot \frac{1}{1 - \frac{1}{j}}$$

$$= \frac{1}{j(j - 1)}$$

$$\sim \frac{1}{j^2}.$$

As $\sum_{j=2}^{\infty} 1/j^2$ converges, it follows by the limit comparison test that

$$\sum_{j=2}^{\infty} \left(\log\left(1 + \frac{i}{j}\right) - \frac{i}{j}\right)$$

is absolutely convergent. Since $\sum_{j=2}^{\infty} i/j$ diverges, we conclude that

$$\sum_{j=2}^{\infty} \log \left(1 + \frac{i}{j} \right)$$

is divergent. The divergence of the given product now follows from Theorem 2.2.4.

An immediate consequence of Theorem 2.2.5 concerns rearrangements of the product factors. Let $\{j_n\}$ be a sequence of natural numbers in which each natural number appears exactly once. Then the product $\prod_{n=1}^{\infty}(1 + z_{j_n})$ is a **rearrangement** of the product $\prod_{j=1}^{\infty}(1 + z_j)$. The next result is the infinite product analogue of Theorem 1.3.6.

Theorem 2.2.9 *All rearrangements of an absolutely convergent product of positive factors converge to the same number.*

Proof Let $\prod_{n=1}^{\infty}(1 + z_{j_n})$ be a rearrangement of the absolutely convergent product $\prod_{j=1}^{\infty}(1 + z_j)$. Since $\prod_{j=1}^{\infty}(1 + |z_j|)$ converges, Theorem 2.2.1 implies that $\sum_{j=1}^{\infty} |z_j|$ converges; hence, the series $\sum_{j=1}^{\infty} \log(1+z_j)$ is absolutely convergent by Theorem 2.2.5. Theorem 1.3.6 shows that all rearrangements of $\sum_{j=1}^{\infty} \log(1 + z_j)$ have the same limit, say L. Therefore $\sum_{n=1}^{\infty} \log(1 + z_{j_n})$ converges to L, and

$$\prod_{n=1}^{\infty}(1 + z_{j_n}) = e^L.$$

□

Exercises 2.2

1. Show that the product

$$\prod_{j=1}^{\infty} j^{1/j}$$

   is divergent. [Hint: Apply Theorem 2.2.1.]
2. Show that the product

$$\prod_{j=1}^{\infty} j^{1/j^2}$$

   is convergent. [Hint: Show that

$$\lim_{x \to 1} \frac{\log x}{x - 1} = 1$$

   and use that equation to show that $j^{1/j^2} - 1$ is of the same order of magnitude as

$$\frac{\log j}{j^2} = \log j^{1/j^2}.]$$

3. Show that the product

$$\prod_{j=1}^{\infty} \left(1 + \frac{(-1)^j}{j+1}\right)$$

converges to 1/2.

4. Show that

$$\prod_{j=1}^{\infty} \left(1 + \frac{(-1)^{j+1}}{\sqrt{j+1}}\right)$$

diverges.

5. Show that

$$\prod_{j=1}^{\infty} \left(1 + \frac{z}{j}\right)$$

is divergent for any $z \neq 0$.

6. Determine whether each of the following products converges:

(a) $\prod_{j=1}^{\infty} j \log \left(1 + \frac{1}{j}\right)$, (Hint: Compare $1 - j \log \left(1 + \frac{1}{j}\right)$ with $1/j$.)

(b) $\prod_{j=1}^{\infty} \cos \frac{1}{j}$. (Hint: $1 - \cos \frac{1}{j} = 2 \sin^2 \frac{1}{2j}$.)

7. Test the convergence of

$$\prod_{j=1}^{\infty} \left(1 - \frac{z^2}{j^2}\right).$$

8. Discuss the convergence of

$$\prod_{j=1}^{\infty} \left(1 + \frac{1}{j^p}\right).$$

9. Prove that

$$\prod_{j=1}^{\infty} (1 + z^{2j-1}) = \frac{1}{1 - z},$$

where $|z| < 1$.

10. Prove that

$$\prod_{j=1}^{\infty}(1+z^j) = \frac{1}{\prod_{j=1}^{\infty}(1-z^{2j-1})},$$

where $|z| < 1$.

11. For what values of x does the product

$$\prod_{j=1}^{\infty}\left(1 + \left(\frac{jx}{j+1}\right)^j\right)$$

converge absolutely?

12. Show that

$$\prod_{j=1}^{\infty}\left(1 + \frac{(-1)^{j-1}}{j^\alpha}\right)$$

converges absolutely if $\alpha > 1$ and diverges to 0 if $0 < \alpha \le 1/2$.

13. Show that the product

$$\prod_{j=1}^{\infty}\frac{x+x^{2j}}{1+x^{2j}}$$

diverges for $x \in [-1, 1)$, but converges for all other $x \in \mathbb{R}$.

2.3 Conditionally Convergent Products

Corollary 2.2.7 shows that if $\sum_{j=1}^{\infty}|z_j|$ converges, where $z_j \ne -1$ for all j, then $\prod_{j=1}^{\infty}(1 + z_j)$ is convergent. However it may be that the series diverges yet the product converges. We begin with a simple example.

Example 2.3.1 Consider the product

$$P = \prod_{j=2}^{\infty}\left(1 + \frac{(-1)^j}{j}\right).$$

The series $\sum_{j=2}^{\infty} 1/j$ is divergent, but

$$P = \left(1 + \frac{1}{2}\right)\left(1 - \frac{1}{3}\right)\left(1 + \frac{1}{4}\right)\left(1 - \frac{1}{5}\right)\cdots$$

$$= \frac{3}{2} \cdot \frac{2}{3} \cdot \frac{5}{4} \cdot \frac{4}{5} \cdots .$$

Thus if we define $P_n = \prod_{j=2}^{n}(1 + (-1)^j/j)$ for all $n \in \mathbb{N}$, then $P_n = 1$ if n is odd and

$$P_n = \frac{n+1}{n} \to 1$$

if n is even. We conclude that P is a conditionally convergent product. Note that the series

$$\sum_{j=2}^{\infty} \frac{(-1)^j}{j}$$

is conditionally convergent.   △

This example suggests that perhaps the convergence of the series $\sum_{j=1}^{\infty} z_j$ is the key ingredient in a sufficient condition for the convergence of $\prod_{j=1}^{\infty}(1 + z_j)$. It turns out that this is not the case: there are products that converge even though the corresponding series diverge and there are products that diverge even though the corresponding series converge.

Example 2.3.2 For each $j > 0$ let

$$z_{2j-1} = \frac{1}{\sqrt{j}}$$

and

$$z_{2j} = -\frac{1}{\sqrt{j}} + \frac{1}{j}.$$

Note that

$$z_{2j-1} + z_{2j} = \frac{1}{j}.$$

Therefore $\sum_{j=1}^{\infty} z_j$ diverges since the harmonic series is divergent. However

$$(1 + z_{2j-1})(1 + z_{2j}) = 1 + \frac{1}{j^{3/2}}.$$

Hence

$$\prod_{j=1}^{\infty}(1+z_j) = \prod_{j=1}^{\infty}\left(1+\frac{1}{j^{3/2}}\right).$$

We conclude that $\prod_{j=1}^{\infty}(1+z_j)$ is convergent by Theorem 2.2.1.   △

Example 2.3.3 Let $z_j = (-1)^j/\sqrt{j}$ for all $j > 0$. Then $\sum_{j=2}^{\infty} z_j$ is a convergent alternating series. However

$$(1+z_{2j-1})(1+z_{2j}) = \left(1-\frac{1}{\sqrt{2j-1}}\right)\left(1+\frac{1}{\sqrt{2j}}\right)$$

$$= 1 - \frac{1+\sqrt{2j}-\sqrt{2j-1}}{\sqrt{2j}(2j-1)}.$$

Note that

$$\frac{1+\sqrt{2j}-\sqrt{2j-1}}{\sqrt{2j}(2j-1)} \sim \frac{1}{2j}.$$

Since $\sum_{j=2}^{\infty} 1/(2j)$ diverges, so does

$$\sum_{j=2}^{\infty} \frac{1+\sqrt{2j}-\sqrt{2j-1}}{\sqrt{2j}(2j-1)}$$

by the limit comparison test. As the terms of this series are positive but less than 1, $\prod_{j=2}^{\infty}(1+z_j)$ diverges by Theorem 2.2.2.   △

The following test is perhaps the most widely used for convergence of a product.

Theorem 2.3.1 (Cauchy's Test)   *Suppose $z_j \neq -1$ for all j and $\sum_{j=1}^{\infty} |z_j|^2$ converges. Then $\sum_{j=1}^{\infty} z_j$ and $\prod_{j=1}^{\infty}(1+z_j)$ either both converge or both diverge.*

Proof Since $z_n \to 0$ as $n \to \infty$, we may assume that $|z_n| \leq 1/2$ for all n. For each n we have

$$\log(1+z_n) = \sum_{j=1}^{\infty}(-1)^{j+1}\frac{z_n^j}{j} = z_n + \sum_{j=2}^{\infty}(-1)^{j+1}\frac{z_n^j}{j},$$

and so

$$|\log(1+z_n) - z_n| = \left|\frac{z_n^2}{2}\sum_{j=2}^{\infty}(-1)^{j+1}\frac{2z_n^{j-2}}{j}\right|$$

$$\leq \frac{|z_n|^2}{2} \sum_{j=2}^{\infty} |z_n|^{j-2}$$

$$= \frac{|z_n|^2}{2(1 - |z_n|)}$$

$$\leq |z_n|^2.$$

As $\sum_{j=1}^{\infty} |z_j|^2$ converges, so does $\sum_{j=1}^{\infty} |\log(1 + z_j) - z_j|$ by the comparison test. Therefore $\sum_{j=1}^{\infty} (\log(1+z_j) - z_j)$ converges, and it follows that $\sum_{j=1}^{\infty} z_j$ converges if and only if $\sum_{j=1}^{\infty} \log(1+z_j)$ converges. The desired conclusion now follows from Theorem 2.2.4.   □

Corollary 2.3.2 *If $z_j \neq -1$ for all j and $\sum_{j=1}^{\infty} z_j$ and $\sum_{j=1}^{\infty} |z_j|^2$ converge then so does $\prod_{j=1}^{\infty} (1 + z_j)$.*

Suppose that $z_j \neq -1$ for all j and that $\sum_{j=1}^{\infty} z_j$ converges. If $\sum_{j=1}^{\infty} |z_j|^2$ also converges, then Corollary 2.3.2 implies that $\prod_{j=1}^{\infty} (1 + z_j)$ converges as well. We now show that the converse holds if each z_j is real.

Theorem 2.3.3 *Let $\{a_j\}$ be a real sequence such that $\sum_{j=1}^{\infty} a_j$ is convergent. If $\prod_{j=1}^{\infty} (1 + a_j)$ converges, then so does $\sum_{j=1}^{\infty} a_j^2$.*

Proof Since $\sum_{j=1}^{\infty} a_j$ converges, we find that $\lim_{j \to \infty} a_j = 0$. Hence we may assume that $a_j > -1$ for all j, so that $1 + a_j > 0$. Moreover, L'Hôpital's rule shows that

$$\lim_{j \to \infty} \frac{a_j - \log(1 + a_j)}{a_j^2} = \lim_{j \to \infty} \frac{1 - \frac{1}{1+a_j}}{2a_j}$$

$$= \lim_{j \to \infty} \frac{a_j}{2a_j + 2a_j^2}$$

$$= \lim_{j \to \infty} \frac{1}{2 + 4a_j}$$

$$= \frac{1}{2}.$$

Therefore we can choose N large enough so that

$$a_j - \log(1 + a_j) > \frac{a_j^2}{4}$$

for all $j \geq N$. For each $n > N$ we then have

$$\sum_{j=N}^{n} a_j - \frac{1}{4} \sum_{j=N}^{n} a_j^2 > \sum_{j=N}^{n} \log(1 + a_j).$$

As $\sum_{j=1}^{\infty} a_j$ converges, if $\sum_{j=1}^{\infty} a_j^2$ were divergent then $\sum_{j=1}^{\infty} \log(1 + a_j)$ would also diverge. Theorem 2.2.4 then gives the contradiction that $\prod_{j=1}^{\infty}(1 + a_j)$ would diverge. We conclude that $\sum_{j=1}^{\infty} a_j^2$ converges.   □

Example 2.3.4 Let

$$a_j = \frac{(-1)^j}{\sqrt{j}}$$

for all j. Then $\sum_{j=1}^{\infty} a_j$ is a convergent alternating series but $\sum_{j=1}^{\infty} a_j^2$ is the harmonic series, which diverges. Theorem 2.3.3 thus implies that $\prod_{j=1}^{\infty}(1 + a_j)$ diverges.   △

We also prove the following related theorem.

Theorem 2.3.4 *Let $\{a_j\}$ be a real sequence. If $\prod_{j=1}^{\infty}(1 + a_j)$ and $\sum_{j=1}^{\infty} a_j^2$ both converge, then so does $\sum_{j=1}^{\infty} a_j$.*

Proof Since $\sum_{j=1}^{\infty} a_j^2$ converges, we have $\lim_{j\to\infty} a_j = 0$. Therefore we may assume that $a_j \in (-1, 1)$ for all j. Hence $1 + a_j > 0$. Accordingly, Theorem 2.2.4 shows that $\sum_{j=1}^{\infty} \log(1 + a_j)$ is convergent.

Now

$$\log(1 + a_j) = \sum_{k=1}^{\infty} (-1)^{k+1} \frac{a_j^k}{k}$$

$$= a_j - a_j^2 \delta_j,$$

where

$$\delta_j = \sum_{k=2}^{\infty} (-1)^k \frac{a_j^{k-2}}{k}$$

for all j. Thus

$$\lim_{j\to\infty} \delta_j = \lim_{j\to\infty} \frac{a_j - \log(1 + a_j)}{a_j^2}$$

$$= \frac{1}{2}$$

as in the proof of Theorem 2.3.3.

Next, choose $\varepsilon > 0$. Using the Cauchy principle, we choose N large enough so that $0 < \delta_j < 1$,

$$\left| \sum_{j=m}^{n} \log(1 + a_j) \right| < \varepsilon$$

and

$$\sum_{j=m}^{n} a_j^2 < \varepsilon$$

whenever $j > N$ and $n \geq m \geq N$. For such m and n it follows that

$$\left| \sum_{j=m}^{n} a_j \right| = \left| \sum_{j=m}^{n} \left(\log(1 + a_j) + \delta_j a_j^2 \right) \right|$$

$$\leq \left| \sum_{j=m}^{n} \log(1 + a_j) \right| + \sum_{j=m}^{n} a_j^2$$

$$< 2\varepsilon.$$

By the Cauchy principle it follows that $\sum_{j=1}^{\infty} a_j$ converges.   □

Example 2.3.5 For all $j \in \mathbb{N}$ define

$$a_j = \begin{cases} \frac{1}{k^2} & \text{if } j = 2k - 1, \\ \frac{1}{k} & \text{if } j = 2k. \end{cases}$$

Thus $\sum_{j=1}^{\infty} a_j$ diverges, since the harmonic series does so. On the other hand, $\sum_{j=1}^{\infty} a_j^2$ converges. Theorem 2.3.4 thus implies that $\prod_{j=1}^{\infty} (1 + a_j)$ diverges.   △

We now revisit the products C and S defined in Example 2.2.7. We use Dirichlet's test (Corollary 1.3.10) to establish conditional convergence. Before we begin the next example, however, we give a lemma that is of interest in its own right and useful for the application of Dirichlet's test.

Lemma 2.3.5 *Let ϕ be any real number that is not an integer multiple of 2π. Then for any $n \in \mathbb{N}$ the following equations hold:*

$$\sum_{j=1}^{n} \cos j\phi = -\frac{1}{2} + \frac{\cos n\phi - \cos((n+1)\phi)}{2(1 - \cos \phi)}, \tag{2.3.1}$$

$$\sum_{j=1}^{n} \sin j\phi = \frac{\sin \phi + \sin n\phi - \sin((n+1)\phi)}{2(1 - \cos \phi)}. \tag{2.3.2}$$

Proof Since

$$e^{ij\phi} = \cos j\phi + i \sin j\phi,$$

we have

$$\sum_{j=1}^{n} \cos j\phi = \text{Re} \left(\sum_{j=1}^{n} e^{ij\phi} \right)$$

and

$$\sum_{j=1}^{n} \sin j\phi = \text{Im} \left(\sum_{j=1}^{n} e^{ij\phi} \right).$$

The number ϕ is not a multiple of 2π and so $e^{i\phi} \neq 1$. We can therefore use the partial sums of the geometric series to get

$$\sum_{j=1}^{n} e^{ij\phi} = \sum_{j=0}^{n} (e^{i\phi})^j - 1$$

$$= \frac{1 - e^{i(n+1)\phi}}{1 - e^{i\phi}} - 1$$

$$= \frac{e^{i\phi} - e^{i(n+1)\phi}}{1 - e^{i\phi}} \cdot \frac{1 - e^{-i\phi}}{1 - e^{-i\phi}}$$

$$= \frac{e^{i\phi} - 1 - e^{i(n+1)\phi} + e^{in\phi}}{2 - e^{i\phi} - e^{-i\phi}}$$

$$= \frac{\cos \phi - 1 - \cos((n+1)\phi) + \cos n\phi + i(\sin \phi - \sin((n+1)\phi) + \sin n\phi)}{2 - 2\cos \phi},$$

and the result follows upon consideration of the real and imaginary parts of this expression.   □

Example 2.3.6 Example 2.2.7 shows that the products $C(\phi, \alpha)$ and $S(\phi, \alpha)$ converge for all $\alpha > 1$ (for certain values of ϕ the products converge to 0). Equipped with Cauchy's test, we now look at the case where $1/2 < \alpha \le 1$.

We use Dirichlet's test (Corollary 1.3.10) with $v_n = 1/n^\alpha$ and $u_n = \cos n\phi$. The sequence $\{v_n\}$ clearly meets the conditions of the test. If ϕ is not a multiple of 2π, Lemma 2.3.5 shows that

$$\left| \sum_{j=1}^{n} u_j \right| \leq \frac{1}{2} + \frac{|\cos n\phi| + |\cos((n+1)\phi)|}{2(1 - \cos \phi)}$$

$$\leq \frac{1}{2} + \frac{1}{\Lambda},$$

where

$$\Lambda = 1 - \cos \phi > 0.$$

The sequence $\{u_n\}$ thus meets the conditions of Dirichlet's test. We conclude that the series $\sum_{j=1}^{\infty} c_j$ converges provided ϕ is not a multiple of 2π. Since $c_j^2 \leq 1/j^{2\alpha}$ and $\alpha > 1/2$ we know that the series $\sum_{j=1}^{\infty} c_j^2$ converges. If ϕ is not a multiple of π, then $c_j > -1$ for all $j \in \mathbb{N}$ and therefore Cauchy's test implies that C converges. If ϕ is an odd multiple of π then $c_j > -1$ for all $j > 1$, as in Example 2.2.7; hence the previous argument shows that

$$\prod_{j=2}^{\infty} \left(1 + \frac{\cos j\phi}{j^{\alpha}} \right)$$

must be convergent for all $\phi \in \mathbb{R}$ and so C converges to 0. If ϕ is an even multiple of π then

$$C(\phi, \alpha) = \prod_{j=1}^{\infty} \left(1 + \frac{1}{j^{\alpha}} \right).$$

For $1/2 < \alpha \leq 1$ the series $\sum_{j=1}^{\infty} 1/j^{\alpha}$ diverges and hence by Theorem 2.2.1 the product diverges.

In a similar manner we can show that S converges. If $\phi = (4m - 1)\pi/2$ for some integer m, then S converges to 0. $\triangle$

We know from Riemann's theorem (Theorem 1.3.5) that, if a conditionally convergent series is rearranged, the resulting series may converge to a different sum or diverge. The situation is much the same for conditionally convergent products. The next example illustrates this comment.

Example 2.3.7 Example 2.3.1 shows that the product

$$P = \prod_{j=2}^{\infty} \left(1 + \frac{(-1)^j}{j} \right)$$

converges conditionally to 1. Consider the following rearrangement R of P:

$$R_1 = \left(1 + \frac{1}{2}\right)\left(1 + \frac{1}{4}\right)\left(1 + \frac{1}{6}\right),$$

$$R_2 = R_1\left(1 - \frac{1}{3}\right),$$

$$R_3 = R_2\left(1 + \frac{1}{8}\right)\left(1 + \frac{1}{10}\right)\left(1 + \frac{1}{12}\right),$$

$$R_4 = R_3\left(1 - \frac{1}{5}\right),$$

$$\vdots$$

The product R is thus formed by taking three factors greater than 1, then a factor less than 1, followed again by three factors greater than 1, and so on. We have

$$R_2 = R_1\left(1 - \frac{1}{3}\right) = \left(1 + \frac{1}{4}\right)\left(1 + \frac{1}{6}\right),$$

$$R_3 = \prod_{j=2}^{6}\left(1 + \frac{1}{2j}\right),$$

$$R_4 = \left(1 - \frac{1}{5}\right)\prod_{j=2}^{6}\left(1 + \frac{1}{2j}\right) = \prod_{j=3}^{6}\left(1 + \frac{1}{2j}\right),$$

and for general n it may be proved readily by induction that

$$R_{2n-1} = \prod_{j=n}^{3n}\left(1 + \frac{1}{2j}\right)$$

and

$$R_{2n} = \prod_{j=n+1}^{3n}\left(1 + \frac{1}{2j}\right).$$

The series $\sum_{j=1}^{\infty} 1/(4j^2)$ converges. The proof of Theorem 2.3.1, together with the comparison test, therefore shows that

$$\sum_{j=1}^{\infty}\left(\log\left(1 + \frac{1}{2j}\right) - \frac{1}{2j}\right)$$

converges. We thus have

$$\lim_{n\to\infty}\left(\log R_{2n-1} - \sum_{j=n}^{3n}\frac{1}{2j}\right) = \lim_{n\to\infty}\sum_{j=n}^{3n}\left(\log\left(1+\frac{1}{2j}\right)-\frac{1}{2j}\right)$$

$$= \lim_{n\to\infty}\left(\sum_{j=1}^{3n}\left(\log\left(1+\frac{1}{2j}\right)-\frac{1}{2j}\right)-\sum_{j=1}^{n-1}\left(\log\left(1+\frac{1}{2j}\right)-\frac{1}{2j}\right)\right)$$

$$= \sum_{j=1}^{\infty}\left(\log\left(1+\frac{1}{2j}\right)-\frac{1}{2j}\right)-\sum_{j=1}^{\infty}\left(\log\left(1+\frac{1}{2j}\right)-\frac{1}{2j}\right)$$

$$= 0. \tag{2.3.3}$$

As $\{1/n\}$ is a decreasing sequence, it follows that

$$\frac{1}{2(j+1)} < \int_{j}^{j+1}\frac{1}{2x}\,dx < \frac{1}{2j}$$

for all j. Hence

$$\int_{n}^{3n}\frac{1}{2x}\,dx < \int_{n}^{3n+1}\frac{1}{2x}\,dx$$

$$= \sum_{j=n}^{3n}\int_{j}^{j+1}\frac{1}{2x}\,dx$$

$$< \sum_{j=n}^{3n}\frac{1}{2j}$$

$$= \frac{1}{2n}+\sum_{j=n+1}^{3n}\frac{1}{2j}$$

$$= \frac{1}{2n}+\sum_{j=n}^{3n-1}\frac{1}{2(j+1)}$$

$$< \frac{1}{2n}+\int_{n}^{3n}\frac{1}{2x}\,dx,$$

so that

$$\frac{1}{2}\log 3 = \int_{n}^{3n}\frac{1}{2x}\,dx < \sum_{j=n}^{3n}\frac{1}{2j} < \frac{1}{2n}+\frac{1}{2}\log 3.$$

We thus have

$$\lim_{n\to\infty} \sum_{j=n}^{3n} \frac{1}{2j} = \frac{1}{2}\log 3 = \log\sqrt{3}.$$

Equation (2.3.3) implies

$$\lim_{n\to\infty} \log R_{2n-1} = \log\sqrt{3};$$

hence,

$$\lim_{n\to\infty} R_{2n-1} = \sqrt{3}.$$

Since

$$R_{2n} = \frac{R_{2n-1}}{1 + \frac{1}{2n}},$$

we also have

$$\lim_{n\to\infty} R_{2n} = \sqrt{3}.$$

Thus the rearranged product converges to $\sqrt{3}$. △

This example illustrates a result that concerns products of the form

$$\prod_{j=1}^{\infty} \left(1 + (-1)^j a_j\right),$$

where $a_j > 0$ for all $j \in \mathbb{N}$ and $\lim_{j\to\infty} j a_j = L$, under rearrangements such as that given in the example. Suppose that the product is conditionally convergent to P, and that it is rearranged so that sequences of factors greater than 1 alternate with sequences of factors less than 1 and the factors greater than 1 and the factors less than 1 both appear in the same order as in the original product. Let μ denote the limit of the ratio of the number of factors in each sequence of factors greater than 1 to the number of factors in the next sequence of factors less than 1. (For the example above, $\mu = 3$.) It can then be shown that the rearranged product converges to $\mu^{L/2} P$. The reader is directed to [14, p. 111], for the details.

Cauchy's test provides a sufficient but not necessary condition for conditional convergence, for if the series $\sum_{j=1}^{\infty} |z_j|$ diverges then so does $\prod_{j=1}^{\infty} (1 + |z_j|)$ by Theorem 2.2.1. The next example shows that a product can converge conditionally even though the two series in Cauchy's theorem diverge.

Example 2.3.8 Let $\{z_j\}$ be the sequence given by

$$z_{2j-1} = -\frac{1}{\sqrt{j+1}}$$

and

$$z_{2j} = \frac{1}{\sqrt{j+1}} + \frac{1}{j+1} + \frac{1}{(j+1)\sqrt{j+1}}.$$

Since

$$z_{2j-1} + z_{2j} = \frac{1}{j+1} + \frac{1}{(j+1)\sqrt{j+1}},$$

it follows that $\sum_{j=1}^{\infty} z_j$ diverges. Therefore $\sum_{j=1}^{\infty} |z_j|$ diverges. So does $\sum_{j=1}^{\infty} |z_j|^2$, since $z_{2j-1}^2 = 1/(j+1)$. However

$$(1 + z_{2j-1})(1 + z_{2j}) = 1 - \frac{1}{(j+1)^2},$$

and it follows that $\prod_{j=1}^{\infty}(1 + z_j)$ is conditionally convergent.   △

Hardy [29] distinguishes two types of convergence for infinite products. For a sequence $\{z_n\}$, the product $\prod_{j=1}^{\infty}(1 + z_j)$ is called **regularly convergent** if either $\sum_{j=1}^{\infty} |z_j|$ converges or both $\sum_{j=1}^{\infty} z_j$ and $\sum_{j=1}^{\infty} |z_j|^2$ converge. Example 2.3.8 shows that there are convergent products that are not regularly convergent. Hardy terms such products **irregularly convergent**. This idea is discussed in more detail in Sect. 2.7.

Pringsheim's extension of Cauchy's test elucidates the rôle of absolute convergence for the series.

Theorem 2.3.6 (Pringsheim's Test) *Let $\{z_n\}$ be a complex sequence, and suppose that $z_n \neq -1$ for all $n \in \mathbb{N}$. Suppose also that there exists $N > 1$ such that the series*

$$\sum_{j=1}^{\infty} z_j, \quad \sum_{j=1}^{\infty} z_j^2, \quad \cdots, \quad \sum_{j=1}^{\infty} z_j^{N-1}, \quad \sum_{j=1}^{\infty} |z_j|^N$$

converge. Then the product $\prod_{j=1}^{\infty}(1 + z_j)$ converges.

Proof We can readily adapt the proof of Cauchy's test to establish this result using the Maclaurin series

$$\log(1 + z) = \sum_{j=1}^{\infty}(-1)^{j+1}\frac{z^j}{j},$$

valid for all z such that $|z| < 1$. Since the series $\sum_{j=1}^{\infty} z_j$ converges, $z_n \to 0$ as $n \to \infty$. We can thus ignore any finite number of factors and assume that $|z_n| < 1/2$ for all n. The condition $z_n \neq -1$ ensures that no factors can vanish. For all j let

$$\sigma_j = \sum_{k=1}^{N-1} (-1)^{k+1} \frac{z_j^k}{k},$$

and

$$\tau_j = \sum_{k=N}^{\infty} (-1)^{k+1} \frac{z_j^k}{k}.$$

Then

$$|\log(1 + z_j) - \sigma_j| = |\tau_j|.$$

By hypothesis the series $\sum_{j=1}^{\infty} |z_j|^N$ converges and since

$$|\tau_j| \leq \sum_{k=N}^{\infty} \frac{|z_j|^k}{k}$$

$$\leq |z_j|^N \sum_{k=0}^{\infty} |z_j|^k$$

$$\leq |z_j|^N \sum_{k=0}^{\infty} \frac{1}{2^k}$$

$$= 2|z_j|^N,$$

the series $\sum_{j=1}^{\infty} (\log(1+z_j) - \sigma_j)$ converges absolutely by the comparison test. For all n let

$$P_n = \prod_{j=1}^{n} (1 + z_j)$$

and

$$Q_n = \log P_n - S_n,$$

where

$$S_n = \sum_{j=1}^{n} \sigma_j = \sum_{k=1}^{N-1} \frac{(-1)^{k+1}}{k} \sum_{j=1}^{n} z_j^k.$$

Thus

$$Q_n = \sum_{j=1}^{n} (\log(1 + z_j) - \sigma_j).$$

We have therefore shown that the sequence $\{Q_n\}$ converges. The condition that the series $\sum_{j=1}^{\infty} z_j^m$ converge for all $m \in \mathbb{N}$ such that $m < N$ ensures that the sequence $\{S_n\}$ also converges. We conclude that the sequence $\{\log P_n\}$ must also converge and therefore the product converges by Theorem 2.2.4.   □

The definition of regular convergence can be extended to cover products that satisfy the conditions of Pringsheim's test. Thus the product $\prod_{j=1}^{\infty} (1 + z_j)$ is **regularly convergent** if $\{z_n\}$ is a complex sequence satisfying the hypotheses of Pringsheim's test. A convergent product is **irregularly convergent** if it is not regularly convergent.

The following example, due to Hardy (*op. cit.*), illustrates the use of Pringsheim's test.

Example 2.3.9 Consider the infinite product

$$P(\phi) = \prod_{j=1}^{\infty} (1 + z_j(\phi)), \tag{2.3.4}$$

where ϕ is real and

$$z_j(\phi) = \frac{e^{ij\phi}}{\sqrt{j}} = \frac{\cos j\phi}{\sqrt{j}} + i \frac{\sin j\phi}{\sqrt{j}}.$$

If ϕ is not a multiple of 2π we can use Dirichlet's test to show that $\sum_{j=1}^{\infty} \mathrm{Re}\,(z_j(\phi))$ and $\sum_{j=1}^{\infty} \mathrm{Im}\,(z_j(\phi))$ converge (see Example 2.3.6). We will assume throughout this example that ϕ is not a multiple of π. Thus $\sum_{j=1}^{\infty} z_j(\phi)$ converges. Since

$$z_j^2(\phi) = \frac{\cos 2j\phi}{j} + i \frac{\sin 2j\phi}{j},$$

we can use Dirichlet's test to show that $\sum_{j=1}^{\infty} z_j^2(\phi)$ also converges. The convergence of these series is conditional, since $|z_j(\phi)| = 1/\sqrt{j}$ and $|z_j(\phi)|^2 = 1/j$. In contrast, $|z_j(\phi)|^3 = 1/j^{3/2}$ and therefore the series $\sum_{j=1}^{\infty} |z_j(\phi)|^3$ converges. Since $(\cos j\phi)/\sqrt{j} > -1$, Pringsheim's test therefore implies that the product $P(\phi)$ converges regularly. The same analysis can be used on the product

$$\bar{P}(\phi) = \prod_{j=1}^{\infty}(1 + \overline{z_j(\phi)}) = \prod_{j=1}^{\infty}\left(1 + \frac{e^{-ij\phi}}{\sqrt{j}}\right)$$

to establish regular convergence.

For each $n \in \mathbb{N}$ let

$$P_n(\phi) = \prod_{j=1}^{n}(1 + z_j(\phi)),$$

$$\bar{P}_n(\phi) = \prod_{j=1}^{n}(1 + \overline{z_j(\phi)}) = \prod_{j=1}^{n}\overline{1 + z_j(\phi)}$$

and

$$Q_n(\phi) = \prod_{j=1}^{n}|1 + z_j(\phi)|^2.$$

Then

$$P_n(\phi)\bar{P}_n(\phi) = Q_n(\phi),$$

and since both $P_n(\phi)$ and $\bar{P}_n(\phi)$ converge, the product

$$\begin{aligned}
Q(\phi) &= \prod_{j=1}^{\infty}\left|1 + \frac{e^{ij\phi}}{\sqrt{j}}\right|^2 \\
&= \prod_{j=1}^{\infty}\left(\left(1 + \frac{\cos j\phi}{\sqrt{j}}\right)^2 + \frac{\sin^2 j\phi}{j}\right) \\
&= \prod_{j=1}^{\infty}\left(1 + \frac{2\cos j\phi}{\sqrt{j}} + \frac{1}{j}\right)
\end{aligned} \tag{2.3.5}$$

converges if ϕ is not a multiple of π. The series $\sum_{j=1}^{\infty}(\cos j\phi)/\sqrt{j}$ converges but $\sum_{j=1}^{\infty} 1/j$ is divergent. We conclude that the series

$$\sum_{j=1}^{\infty}\left(\frac{2\cos j\phi}{\sqrt{j}} + \frac{1}{j}\right)$$

diverges. The product (2.3.5) is therefore irregularly convergent even though the products $P(\phi)$ and $\bar{P}(\phi)$ are regularly convergent. $\triangle$

Hardy's work brought to the fore the product

$$\prod_{j=2}^{\infty} \left(1 + \frac{e^{ij\phi}}{\log j} \right). \tag{2.3.6}$$

He proved that the product diverges if ϕ/π is rational. Littlewood [37] showed that there is a class of irrational values of ϕ/π for which the product converges. Now let

$$z_j = \frac{e^{ij\phi}}{\log j}$$

for each $j > 1$. For any $N \in \mathbb{N}$,

$$z_j^N = \frac{e^{iNj\phi}}{\log^N j}.$$

Dirichlet's test can be used to show that $\sum_{j=2}^{\infty} z_j^N$ converges whenever ϕ/π is irrational. On the other hand,

$$|z_j|^N = \frac{1}{\log^N j},$$

and since

$$\log^N j << j,$$

for any $N \in \mathbb{N}$ we see that $\sum_{j=2}^{\infty} |z_j|^N$ diverges. Thus if the product converges then it does so irregularly.

A product $\prod_{j=1}^{\infty} (1 + z_j)$ may be irregularly convergent because it converges but at least one of the series

$$\sum_{j=1}^{\infty} z_j, \quad \sum_{j=1}^{\infty} z_j^2, \quad \cdots, \quad \sum_{j=1}^{\infty} z_j^{N-1}$$

in Pringsheim's test diverges. This work of Hardy and Littlewood is interesting because it shows that there are irregularly convergent products such that $\sum_{j=2}^{\infty} z_j^N$ converges for any $N \in \mathbb{N}$ but never absolutely.

Littlewood proved something more general than the convergence of (2.3.6) for certain values of ϕ. He studied products of the form

$$\prod_{j=1}^{\infty} \left(1 + z_j e^{ij\phi} \right), \tag{2.3.7}$$

where $z_j \to 0$ as $j \to \infty$. He showed that there is a class of irrational values of ϕ/π for which the product converges, and that this class of values is independent of the sequence $\{z_n\}$.

Exercises 2.3

1. Let $\{a_j\}$ be a sequence that decreases monotonically to 0. Show that $\sum_{j=1}^{\infty} a_j^2$
   converges if and only if $\prod_{j=1}^{\infty}(1 + (-1)^j a_j)$ does so.
2. Find the values of α for which

$$\prod_{j=1}^{\infty}\left(1 + \frac{(-1)^j}{j^\alpha}\right)$$

   is convergent.
3. Show that if $\prod_{j=1}^{\infty}(1+a_j)$ and $\prod_{j=1}^{\infty}(1-a_j)$ both converge, then so do $\sum_{j=1}^{\infty} a_j$
   and $\sum_{j=1}^{\infty} a_j^2$.
4. Show that the product

$$\prod_{j=2}^{\infty}\left(1 + \frac{(-1)^j}{j^{2/3}}\right)$$

   converges conditionally.
5. Show that

$$\prod_{j=1}^{\infty}\left(1 + \frac{(-1)^{j-1}}{j^\alpha}\right)$$

   converges conditionally if $1/2 < \alpha \le 1$.
6. Let P be the product of Example 2.3.7 and consider the rearrangement of P to a product R given by

$$R_1 = \left(1 + \frac{1}{2}\right)\left(1 + \frac{1}{4}\right)\left(1 + \frac{1}{6}\right)\left(1 + \frac{1}{8}\right),$$

$$R_2 = R_1\left(1 - \frac{1}{3}\right),$$

$$R_3 = R_2\left(1 + \frac{1}{10}\right)\left(1 + \frac{1}{12}\right)\left(1 + \frac{1}{14}\right)\left(1 + \frac{1}{16}\right)$$

$$R_4 = R_3\left(1 - \frac{1}{5}\right)$$

$$\vdots$$

What is the value of R?

7. Show that for all $x \in \mathbb{R}$,

$$\lim_{n \to \infty} \prod_{j=n+1}^{2n} \left(1 - \frac{x}{j}\right) = 2^{-x}.$$

8. Let $\{a_n\}$ be the sequence defined by

$$a_{2j} = \frac{1}{\sqrt{j} - \frac{1}{2}}, \quad a_{2j+1} = -\frac{1}{\sqrt{j} + \frac{1}{2}}.$$

Show that the product $\prod_{j=1}^{\infty}(1 + a_j)$ converges irregularly.

2.4   Uniform Convergence of Products of Functions

Infinite products that define analytic functions are of central interest in analysis. In this section we combine the uniform convergence of a product of analytic functions with Theorem 1.5.1 to obtain conditions under which a product defines an analytic function. Our first result is an easy consequence of Theorem 1.5.1.

Theorem 2.4.1 *Let $\{f_n\}$ be a sequence of functions analytic in some domain Ω. Suppose that $\prod_{j=1}^{\infty} f_j(z)$ converges uniformly to f on every compact subset of Ω. Then f is analytic on Ω.*

Proof Let D be a compact subset of Ω. Thus $\prod_{j=1}^{\infty} f_j(z)$ converges uniformly to f on D. The sequence $\{\prod_{j=1}^{n} f_j(z)\}$ therefore converges uniformly to f on D. Each term of this sequence is a finite product and hence analytic. Thus f is analytic by Theorem 1.5.1.   □

The next two theorems provide sufficient conditions for the uniform convergence of a product.

Theorem 2.4.2 *Let $\{f_n\}$ be a sequence of bounded functions defined on a set Ω. Suppose that $f_n(z) \neq 0$ for all $n \in \mathbb{N}$ and $z \in \Omega$. If $\sum_{j=1}^{\infty} \log f_j(z)$ converges uniformly on Ω to a bounded function, then so does $\prod_{j=1}^{\infty} f_j(z)$.*

Proof Note first that the exponential function is uniformly continuous on any compact subset of $\mathbb{C}$. Theorem 1.4.8 therefore shows that

$$\prod_{j=1}^{n} f_j(z) = \prod_{j=1}^{n} \exp(\log f_j(z))$$

$$= \exp\left(\sum_{j=1}^{n} \log f_j(z)\right)$$

$$\to \exp\left(\sum_{j=1}^{\infty} \log f_j(z)\right)$$

$$\neq 0$$

as $n \to \infty$. Moreover the convergence is uniform and to a bounded function. □

Theorems 2.4.2 and 2.4.1 immediately give the following corollary.

Corollary 2.4.3 *Let $\{f_n\}$ be a sequence of bounded analytic functions on some domain Ω. Suppose that $f_n(z) \neq 0$ for all $n \in \mathbb{N}$ and $z \in \Omega$. If $\sum_{j=1}^{\infty} \log f_j(z)$ converges uniformly on Ω to a bounded function, then $\prod_{j=1}^{\infty} f_j(z)$ is analytic on Ω.*

Theorem 2.4.4 *Let $\{f_n\}$ be a sequence of functions defined and bounded on some set Ω. Suppose that $\sum_{j=1}^{\infty} |f_j(z)|$ converges uniformly on Ω to a bounded function. Then the product $\prod_{j=1}^{\infty}(1 + f_j(z))$ converges uniformly on Ω to a bounded function f. Moreover $f(z) = 0$ for some $z \in \Omega$ if and only if $1 + f_n(z) = 0$ for some n.*

Proof The uniform convergence of the series on Ω implies the uniform convergence of $\{|f_n(z)|\}$ to 0 on Ω (Theorem 1.4.15). It follows that there exists L such that $|f_n(z)| \leq 1/2$ for all $z \in \Omega$ and all $n \geq L$. Now

$$\log(1 + f_n(z)) = \sum_{j=1}^{\infty}(-1)^{j+1} \frac{f_n^j(z)}{j}$$

$$= f_n(z)S(f_n(z)),$$

where

$$S(f_n(z)) = \sum_{j=1}^{\infty}(-1)^{j+1} \frac{f_n^{j-1}(z)}{j}$$

$$= \sum_{j=0}^{\infty}(-1)^j \frac{f_n^j(z)}{j+1}.$$

Thus

$$|S(f_n(z))| \leq \sum_{j=0}^{\infty} |f_n^j(z)| \leq 2$$

whenever $n \geq L$.

For any fixed $\varepsilon > 0$, the Cauchy principle guarantees the existence of M such that $\sum_{j=p}^{q} |f_j(z)| < \varepsilon$ whenever $q \geq p \geq M$. For $q \geq p \geq N$, where $N = \max\{L, M\}$, we therefore find that

$$\sum_{j=p}^{q} |\log(1 + f_j(z))| = \sum_{j=p}^{q} |f_j(z)||S(f_j(z))| \tag{2.4.1}$$

$$< 2\varepsilon.$$

Therefore

$$\sum_{j=N}^{\infty} \log(1 + f_j(z))$$

converges uniformly on Ω, by the Cauchy principle and Theorem 1.4.16.

Next, for all $n \geq N$ we write

$$\prod_{j=1}^{n}(1 + f_j(z)) = \prod_{j=1}^{N-1}(1 + f_j(z)) \prod_{j=N}^{n}(1 + f_j(z))$$

and note that all these products give bounded functions. Since $\sum_{j=N}^{\infty} |f_j(z)|$ converges to a bounded function and $|S(f_j(z))| \leq 2$ whenever $j \geq N$, it follows that

$$\sum_{j=N}^{\infty} |f_j(z)||S(f_j(z))|$$

is bounded and consequently so is

$$\sum_{j=N}^{\infty} \log(1 + f_j(z)),$$

by Eq. (2.4.1). The uniform convergence of the product to a bounded function now follows from Theorem 2.4.2.

Finally, $1 + f_n(z) \neq 0$ on Ω for all $n \geq N$. Since

$$f(z) = \prod_{j=1}^{\infty}(1 + f_j(z))$$

for all $z \in \Omega$, it follows that $f(z) = 0$ for some $z \in \Omega$ if and only if $1 + f_n(z) = 0$ for some $n < N$. $\qquad\square$

Theorems 2.4.4 and 2.4.1 immediately give the following corollary, for if f is an analytic function on some domain Ω, then so is the function given by $1 + f(z)$ for all $z \in \Omega$.

Corollary 2.4.5 *Let $\{f_n\}$ be a sequence of bounded analytic functions on some domain Ω. Suppose that $\sum_{j=1}^{\infty} |f_j(z)|$ converges uniformly on Ω to a bounded function. Then $\prod_{j=1}^{\infty}(1 + f_j(z))$ is analytic on Ω.*

Exercises 2.4

1. Let

$$f(z) = \prod_{j=1}^{\infty}(1 + c^j z),$$

where $|c| < 1$. Show that f defines an entire function.

2. Let

$$g(z) = \prod_{j=2}^{n}\left(1 + \frac{z}{j \log^2 j}\right).$$

Show that g defines an entire function.

2.5 Infinite Products of Real Functions

The next result follows from Theorem 2.4.4 and provides a useful test for uniform convergence.

Theorem 2.5.1 *Let $\{f_n\}$ be a sequence of bounded functions defined on an interval I, and suppose that $f_n(x) > -1$ for all $x \in I$ and $n \in \mathbb{N}$. If the series $\sum_{j=1}^{\infty} |f_j(x)|$ is uniformly convergent on I to a bounded function, then $\prod_{j=1}^{\infty}(1 + f_j(x))$ is uniformly convergent on I to a bounded function f such that $f(x) \neq 0$ for all $x \in I$.*

Example 2.5.1 Show that the product

$$P(x) = \prod_{j=2}^{\infty} \cos\frac{x}{2^j} \tag{2.5.1}$$

is uniformly convergent on $I = [-\pi, \pi]$.

Solution The Maclaurin series for $\cos(x/2^j)$ is

$$\cos\frac{x}{2^j} = 1 + f_j(x),$$

where

$$f_j(x) = \sum_{k=1}^{\infty} \frac{(-1)^k x^{2k}}{(2k)! 2^{2kj}}. \tag{2.5.2}$$

We now check that the series (2.5.2) defining $f_j(x)$ satisfies the hypotheses of Leibniz's test (Theorem 1.3.7) for all $x \in I - \{0\}$. Since

$$\frac{x^{2k+2}}{(2k+2)! 2^{2j(k+1)}} \cdot \frac{(2k)! 2^{2kj}}{x^{2k}} = \frac{x^2}{(k+1)(2k+1) 2^{2j+1}}$$

$$\leq \frac{\pi^2}{2 \cdot 3 \cdot 2^3}$$

$$< 1$$

for all $j \in \mathbb{N}$, the sequence

$$\left\{ \frac{x^{2n}}{(2n)! 2^{2nj}} \right\}$$

of positive terms is decreasing for all $j \in \mathbb{N}$. It is constant if $x = 0$. In any case it converges to 0, being a subsequence of

$$\left\{ \frac{1}{n!} \left(\frac{x}{2^j} \right)^n \right\}.$$

Observe that

$$\sum_{k=0}^{\infty} (-1)^k b_k = S$$

where

$$b_k = \frac{x^{2k}}{(2k)! 2^{2kj}},$$

for all k, and

$$S = 1 + f_j(x).$$

Theorem 1.3.8 therefore shows that

$$|S - b_0| \leq b_1,$$

so that

$$-b_1 \leq S - b_0 = S - 1 = f_j(x).$$

Hence

$$f_j(x) \geq -\frac{x^2}{2 \cdot 2^{2j}}.$$

On the other hand,

$$-f_j(x) = \sum_{k=0}^{\infty}(-1)^k c_k$$

where $c_k = b_{k+1}$. Therefore

$$c_1 \geq |-f_j(x) - c_0| = |f_j(x) + c_0|,$$

so that

$$f_j(x) \leq -c_0 + c_1 = -b_1 + b_2 \leq 0.$$

We conclude that

$$|f_j(x)| \leq \frac{x^2}{2 \cdot 2^{2j}} \leq \frac{\pi^2}{2} \cdot \frac{1}{4^j}. \tag{2.5.3}$$

The series $\sum_{j=2}^{\infty} 1/4^j$ converges and from the Weierstrass M-test we deduce that $\sum_{j=2}^{\infty} |f_j(x)|$ is uniformly convergent on I to a bounded function. Inequality (2.5.3) shows that $f_j(x) > -1$ for all $j > 1$ and $x \in I$. We conclude from Theorem 2.5.1 that the product is uniformly convergent on I.   △

Example 2.5.2 The example above can be used to derive an infinite product representation for $\sin x$ and another formula for π.

For all $x \in \mathbb{R}$, we have the double angle relation

$$\sin x = 2 \sin \frac{x}{2} \cos \frac{x}{2}. \tag{2.5.4}$$

Equation (2.5.4) can be applied to $\sin(x/2)$ to get

$$\sin \frac{x}{2} = 2 \sin \frac{x}{2^2} \cos \frac{x}{2^2}$$

and combined with Eq. (2.5.4) this equation yields

$$\sin x = 2^2 \sin \frac{x}{2^2} \cos \frac{x}{2} \cos \frac{x}{2^2}.$$

Evidently we can continue this process inductively to get

$$\sin x = 2^n \sin \frac{x}{2^n} \prod_{j=1}^{n} \cos \frac{x}{2^j}. \tag{2.5.5}$$

Using L'Hôpital's rule we see that

$$
\begin{aligned}
\lim_{n\to\infty} 2^n \sin \frac{x}{2^n} &= \lim_{n\to\infty} \frac{\sin \frac{x}{2^n}}{\frac{1}{2^n}} \\
&= \lim_{n\to\infty} \frac{-\frac{x\log 2}{2^n} \cos \frac{x}{2^n}}{-\frac{\log 2}{2^n}} \\
&= \lim_{n\to\infty} x \cos \frac{x}{2^n} \\
&= x.
\end{aligned}
$$

We thus have

$$\sin x = x \prod_{j=1}^{\infty} \cos \frac{x}{2^j} \tag{2.5.6}$$

for all $x \in [-\pi, \pi]$.

Example 2.5.1 shows that the infinite product (2.5.1) is uniformly convergent on $[-\pi, \pi]$. In fact, similar arguments can be used to show that the infinite product $\prod_{j=1}^{\infty} \cos(x/2^j)$ converges uniformly in any closed interval I that does not contain a nonzero integer multiple of π. Moreover the product converges to zero when x is an odd multiple of π, because $\cos((2n+1)\pi/2) = 0$ for any integer n. If x is a nonzero even multiple of π, we can write $x = 2^n m\pi$ where $n \in \mathbb{N}$ and m is odd. Then

$$\cos \frac{2^n m\pi}{2^{n+1}} = 0,$$

and again the product converges to 0. We conclude that Eq. (2.5.6) is valid for all $x \in \mathbb{R}$.

For each $x \in (0, \pi)$, Eq. (2.5.6) can be recast as

$$\frac{\sin x}{x \cos \frac{x}{2}} = \prod_{j=2}^{\infty} \cos \frac{x}{2^j},$$

and L'Hôpital's rule implies

$$\prod_{j=2}^{\infty} \cos \frac{\pi}{2^j} = \lim_{x \to \pi^-} \frac{\sin x}{x \cos \frac{x}{2}} = \lim_{x \to \pi^-} \frac{\cos x}{\cos \frac{x}{2} - \frac{x}{2} \sin \frac{x}{2}} = \frac{2}{\pi}.$$

The double angle relation

$$\cos^2 x = \frac{1 + \cos 2x}{2}$$

gives

$$\cos \frac{x}{2} = \frac{1}{2}\sqrt{2 + 2\cos x}$$

for all $x \in [-\pi, \pi]$; hence,

$$\cos \frac{\pi}{2^2} = \frac{1}{2}\sqrt{2 + 2\cos \frac{\pi}{2}} = \frac{\sqrt{2}}{2},$$

$$\cos \frac{\pi}{2^3} = \frac{1}{2}\sqrt{2 + 2\cos \frac{\pi}{2^2}} = \frac{\sqrt{2 + \sqrt{2}}}{2},$$

$$\cos \frac{\pi}{2^4} = \frac{1}{2}\sqrt{2 + 2\cos \frac{\pi}{2^3}} = \frac{\sqrt{2 + \sqrt{2 + \sqrt{2}}}}{2},$$

and, in general, if $b_n = 2\cos(\pi/2^n)$ then $b_1 = 0$ and

$$b_{n+1} = \sqrt{2 + b_n}$$

for all $n \in \mathbb{N}$. We thus get Viète's formula:

$$\frac{2}{\pi} = \frac{\sqrt{2}}{2} \cdot \frac{\sqrt{2 + \sqrt{2}}}{2} \cdot \frac{\sqrt{2 + \sqrt{2 + \sqrt{2}}}}{2} \cdots. \tag{2.5.7}$$

This late sixteenth century discovery gives the earliest expression for π in terms of an infinite product. Although it may not provide the most efficient method for approximating π in terms of partial products, it is certainly a striking relation: the number π is expressed purely in terms of the number 2 and square roots.   △

The results of Sect. 1.4 can be applied directly to uniformly convergent products.

Theorem 2.5.2 *Let $\{f_n\}$ be a sequence of real functions defined on an interval I and suppose that $f_n(x) \neq -1$ for all $x \in I$ and all $n \in \mathbb{N}$. Suppose also that the product $\prod_{j=1}^{\infty}(1 + f_j(x))$ is uniformly convergent to P on I.*

1. If f_n is continuous on I for all $n \in \mathbb{N}$, then P is continuous on I.

2. If f_n is integrable on I for all $n \in \mathbb{N}$, then P is integrable on I.

Proof If f and g are functions that are continuous (integrable) over I, then $f + g$ and fg are continuous (integrable) over I. Each member of the sequence $\{P_n\}$ of partial products is thus continuous (integrable) over I. The result follows immediately from Corollary 1.4.10 and Theorem 1.4.11. $\square$

Given a sequence $\{f_n\}$ of real differentiable functions, the partial products

$$P_n = \prod_{j=1}^{n}(1 + f_j(x))$$

are differentiable. Theorem 1.4.14, however, requires that the sequence $\{P_n'\}$ be uniformly convergent. The next result gives conditions under which this sequence is uniformly convergent and provides a formula for the derivative of an infinite product.

Theorem 2.5.3 *Let $\{f_n\}$ be a sequence of real functions that have continuous derivatives on a closed interval I, and suppose that:*

1. $f_n(x) > -1$ for all $x \in I$ and $n \in \mathbb{N}$;
2. there is a $c \in I$ such that $\prod_{j=1}^{\infty}(1 + f_j(c))$ converges; and
3. the series

$$\sum_{j=1}^{\infty} \frac{f_j'(x)}{1 + f_j(x)}$$

is uniformly convergent on I.

Then the product $\prod_{j=1}^{\infty}(1 + f_j(x))$ is uniformly convergent on I to a differentiable function P and, for all $x \in I$,

$$P'(x) = \left(\prod_{j=1}^{\infty}(1 + f_j(x))\right) \sum_{j=1}^{\infty} \frac{f_j'(x)}{1 + f_j(x)}. \qquad (2.5.8)$$

Proof Let

$$P_n(x) = \prod_{j=1}^{n}(1 + f_j(x))$$

for all $n \in \mathbb{N}$ and $x \in I$. Note that P_n is differentiable, and hence continuous and bounded, on the closed interval I for each n. Furthermore,

$$\log P_n(x) = \sum_{j=1}^{n} \log(1 + f_j(x)),$$

and since f_j is differentiable,

$$P_n'(x) = P_n(x)\sigma_n(x),$$

where

$$\sigma_n(x) = \sum_{j=1}^{n} \frac{f_j'(x)}{1 + f_j(x)}.$$

For each $j \in \mathbb{N}$, the function $f_j'(x)/(1 + f_j(x))$ is continuous and therefore bounded on I, so that σ_n is bounded on I. By hypothesis, $\{\sigma_n\}$ is a uniformly convergent sequence. Since $\prod_{j=1}^{\infty}(1 + f_j(c))$ converges for some $c \in I$ we see that the series $\sum_{j=1}^{\infty} \log(1 + f_j(c))$ converges. Theorem 1.4.14 thus implies that $\{\log P_n\}$ converges uniformly on I to a function which is differentiable, and therefore continuous and bounded, on the closed interval I. As $P_n = e^{\log P_n}$ for all n, Theorem 1.4.8 therefore shows that $\{P_n\}$ converges uniformly on I to a function P. Since $\{\sigma_n\}$ and $\{P_n\}$ converge uniformly on I and are bounded, Corollary 1.4.7 implies that the sequence $\{P_n \sigma_n\} = \{P_n'\}$ converges uniformly on I. From Theorem 1.4.14 we therefore infer that P is differentiable on I and that P' is given by Eq. (2.5.8). $\square$

Example 2.5.3 Let

$$P(x) = \prod_{j=1}^{\infty}(1 + x^j) = \prod_{j=1}^{\infty}(1 + f_j(x))$$

where $f_j(x) = x^j$ for all $x \in \mathbb{R}$. The series $\sum_{j=1}^{\infty} |x|^j$ is uniformly convergent on any closed interval $I = [a, b]$ included in $(-1, 1)$ (since $\sum_{j=1}^{\infty} x^j$ is a power series with a unit radius of convergence), and it converges to a function that is continuous and hence bounded. Theorem 2.5.1 thus implies that P is uniformly convergent on I. For each $j \in \mathbb{N}$ and $x \in I$,

$$\left| \frac{f_j'(x)}{1 + f_j(x)} \right| = \left| \frac{jx^{j-1}}{1 + x^j} \right|$$

$$\leq \frac{j|x|^{j-1}}{1 - |x|^j}$$

$$\leq \frac{j\Lambda^{j-1}}{1 - \Lambda^j},$$

where $\Lambda = \max\{|a|, |b|\}$. Since $0 \le \Lambda < 1$, $\Lambda^j \to 0$ as $j \to 0$ and therefore there is an N such that $\Lambda^j \le 1/2$ for all $j \ge N$. We thus have, for all $x \in I$ and $j \ge N$,

$$\left| \frac{f_j'(x)}{1 + f_j(x)} \right| \le 2j\Lambda^{j-1}.$$

Since $\sum_{j=1}^{\infty} j\Lambda^{j-1}$ converges by the ratio test, the series

$$\sum_{j=1}^{\infty} \frac{jx^{j-1}}{1 + x^j}$$

converges uniformly on I by the Weierstrass M-test. In addition, $x^j > -1$ for all $x \in I$ and $j \in \mathbb{N}$; therefore, the conditions of Theorem 2.5.3 are satisfied. We thus have

$$P'(x) = \left(\prod_{j=1}^{\infty} (1 + x^j) \right) \sum_{j=1}^{\infty} \frac{jx^{j-1}}{1 + x^j}$$

for all $x \in I$.   △

Theorem 2.5.1 requires that the series $\sum_{j=1}^{\infty} |f_j(x)|$ be uniformly convergent on I. It may be, however, that $\sum_{j=1}^{\infty} f_j(x)$ is not absolutely convergent at some $x \in I$ yet the product $\prod_{j=1}^{\infty} (1 + f_j(x))$ nonetheless converges uniformly on I. The next theorem is a straightforward adaptation of Theorem 2.3.1. The proof is left as an exercise.

Theorem 2.5.4 (Cauchy's Test) *Suppose that $\sum_{j=1}^{\infty} f_j(x)$ and $\sum_{j=1}^{\infty} f_j^2(x)$ are uniformly convergent to bounded functions on an interval I. Furthermore, suppose that $f_j(x) > -1$ for all $x \in I$ and $j \in \mathbb{N}$. Then the product $\prod_{j=1}^{\infty} (1 + f_j(x))$ converges uniformly on I.*

Example 2.5.4 Let $I = [a, b]$, where $0 < a < b < 2\pi$, and, for each $x \in I$, let

$$P(x) = \prod_{j=2}^{\infty} \left(1 + f_j(x) \right),$$

where

$$f_j(x) = \frac{\sin jx}{j}.$$

If $a \le \pi/2 \le b$ then the series $\sum_{j=2}^{\infty} f_j(x)$ is not absolutely convergent for all $x \in I$, but it is uniformly convergent to a function that is continuous, and therefore bounded, on the closed interval I (Example 2.3.6). For all $x \in I$,

$$f_j^2(x) = \frac{\sin^2 jx}{j^2} \le \frac{1}{j^2},$$

and since the series $\sum_{j=2}^{\infty} 1/j^2$ converges, the Weierstrass M-test implies that the series $\sum_{j=2}^{\infty} f_j^2(x)$ is uniformly convergent on I to a bounded function. Note also that $-1/2 \le f_j(x) \le 1/2$ for all $x \in I$ and $j \ge 2$. Theorem 2.5.4 thus shows that the product is uniformly convergent on I. △

The results of this section can be readily generalized to accommodate sequences of complex functions. In particular, Pringsheim's test can be readily adapted for uniform convergence. The proof is left as an exercise.

Theorem 2.5.5 *Let* $\{f_n(z)\}$ *be a sequence of complex-valued functions defined on the closed disc*

$$\overline{N}_r(z_0) = \{z \in \mathbb{C} : |z - z_0| \le r\}$$

and suppose that $f_n(z) \ne -1$ *for all* $n \in \mathbb{N}$ *and* $z \in \overline{N}_r(z_0)$. *Suppose also that there is an* $N \in \mathbb{N}$ *such that the series*

$$\sum_{j=1}^{\infty} f_j(z), \quad \sum_{j=1}^{\infty} f_j^2(z), \quad \ldots, \quad \sum_{j=1}^{\infty} f_j^{N-1}(z), \quad \sum_{j=1}^{\infty} |f_j(z)|^N$$

converge uniformly on $\overline{N}_r(z_0)$ *to bounded functions. Then the product*

$$\prod_{j=1}^{\infty} (1 + f_j(z))$$

converges uniformly on $\overline{N}_r(z_0)$.

Exercises 2.5

1. Let $\{a_n\}$ be a sequence of numbers such that $\sum_{j=1}^{\infty} a_j$ converges absolutely. Show that the product

$$\prod_{j=1}^{\infty} (1 + a_j x^j)$$

is uniformly convergent on $[0, 1]$ and that

$$\lim_{x \to 1^-} \prod_{j=1}^{\infty} (1 + a_j x^j) = \prod_{j=1}^{\infty} (1 + a_j).$$

2. Show that the product

$$\prod_{j=1}^{\infty} \left(1 + \sin^2 \frac{x}{2^j}\right)$$

converges uniformly on $[-\pi, \pi]$.
3. Prove Theorems 2.5.4 and 2.5.5.
4. If $\alpha > 1/2$, show that the product

$$\prod_{j=2}^{\infty} \left(1 + \frac{\sin jx}{j^\alpha}\right)$$

converges uniformly on $[\pi/4, \pi/2]$.

2.6   Infinite Product Expansions for $\sin x$ and $\cos x$

Infinite product expansions of functions provide useful tools in analysis and are particularly suited to problems where the zeros of a function are of central interest. In this section we derive infinite product representations for $\sin x$ and $\cos x$ using elementary principles as opposed to methods that involve complex analysis. Our derivation is based on that given by Venkatachaliengar [68]. We state and prove the result for a real variable. The proof can be adapted for a complex variable and the expansion is formally the same (cf. Venkatachaliengar *op. cit.*).

Theorem 2.6.1   *For all $x \in \mathbb{R}$:*

$$\sin x = x \prod_{j=1}^{\infty} \left(1 - \frac{x^2}{j^2 \pi^2}\right), \tag{2.6.1}$$

$$\cos x = \prod_{j=1}^{\infty} \left(1 - \frac{4x^2}{(2j-1)^2 \pi^2}\right). \tag{2.6.2}$$

Remark If x is an integer multiple of π, the infinite product in Eq. (2.6.1) converges to zero in the sense described in Sect. 2.1. A similar remark holds for the product that represents $\cos x$.

Proof Define the sequence $\{I_n\}$ of functions by

$$I_n(x) = \int_0^{\pi/2} \cos x\xi \, \cos^n \xi \, d\xi$$

for all $x \in \mathbb{R}$ and $n \geq 0$. If $n \geq 2$ and $x \neq 0$, then integration by parts gives

$$I_n(x) = \left. \frac{\cos^n \xi \sin x\xi}{x} \right|_0^{\pi/2} + \frac{n}{x} \int_0^{\pi/2} \cos^{n-1} \xi \sin \xi \sin x\xi \, d\xi$$

$$= \frac{n}{x} \int_0^{\pi/2} \cos^{n-1} \xi \sin \xi \sin x\xi \, d\xi.$$

Integrating by parts again yields

$$x I_n(x) = - \left. \frac{n \cos^{n-1} \xi \sin \xi \cos x\xi}{x} \right|_0^{\pi/2}$$

$$+ \frac{n}{x} \int_0^{\pi/2} \left(-(n-1) \cos^{n-2} \xi \sin^2 \xi + \cos^n \xi \right) \cos x\xi \, d\xi$$

$$= \frac{n}{x} \int_0^{\pi/2} \left(-(n-1)(1 - \cos^2 \xi) \cos x\xi \cos^{n-2} \xi + \cos x\xi \cos^n \xi \right) d\xi$$

$$= \frac{n}{x} \left(-(n-1) \int_0^{\pi/2} \cos x\xi \cos^{n-2} \xi \, d\xi + n \int_0^{\pi/2} \cos x\xi \cos^n \xi \, d\xi \right)$$

$$= \frac{n}{x} (-(n-1) I_{n-2}(x) + n I_n(x)).$$

Therefore, if $x \neq 0$ and $n \geq 2$,

$$(n^2 - x^2) I_n(x) - n(n-1) I_{n-2}(x) = 0. \tag{2.6.3}$$

If $x = 0$ and $n \geq 2$ then

$$I_n(0) = \int_0^{\pi/2} \cos^n \xi \, d\xi = \frac{n-1}{n} I_{n-2}(0),$$

so that

$$n^2 I_n(0) - n(n-1) I_{n-2}(0) = 0.$$

Thus Eq. (2.6.3) is satisfied for all $x \in \mathbb{R}$ whenever $n \geq 2$. If $x \neq 0$,

$$I_0(x) = \int_0^{\pi/2} \cos x\xi \, d\xi = \frac{1}{x} \sin \frac{\pi x}{2},$$

but

$$I_0(0) = \frac{\pi}{2}.$$

For all x, integration by parts twice gives

$$I_1(x) = \int_0^{\pi/2} \cos x\xi \cos \xi \, d\xi$$

$$= \sin \xi \cos x\xi \, |_0^{\pi/2} + x \int_0^{\pi/2} \sin \xi \sin x\xi \, d\xi$$

$$= \cos \frac{\pi x}{2} + x \left(-\cos \xi \sin x\xi \, |_0^{\pi/2} + x \int_0^{\pi/2} \cos \xi \cos x\xi \, d\xi \right)$$

$$= \cos \frac{\pi x}{2} + x^2 I_1(x),$$

so that

$$(1 - x^2) I_1(x) = \cos \frac{\pi x}{2}.$$

Hence $I_1(0) = 1$.

We thus have the following equations:

$$\sin \frac{\pi x}{2} = \frac{\pi x}{2} \cdot \frac{I_0(x)}{I_0(0)}, \tag{2.6.4}$$

$$\cos \frac{\pi x}{2} = (1 - x^2) \frac{I_1(x)}{I_1(0)}. \tag{2.6.5}$$

Since $\cos \xi > 0$ for all $\xi \in (0, \pi/2)$, we have $I_n(0) > 0$ for all non-negative integers n. Equation (2.6.3) can be recast as

$$I_{n-2}(x) = \frac{n^2 - x^2}{n(n-1)} I_n(x),$$

which implies that

$$I_{n-2}(0) = \frac{n^2}{n(n-1)} I_n(0),$$

and consequently

$$\frac{I_{n-2}(x)}{I_{n-2}(0)} = \left(1 - \frac{x^2}{n^2} \right) \frac{I_n(x)}{I_n(0)} \tag{2.6.6}$$

for all $n \geq 2$ and $x \in \mathbb{R}$. Equations (2.6.4) and (2.6.6) imply

$$\sin \frac{\pi x}{2} = \frac{\pi x}{2} \cdot \frac{I_0(x)}{I_0(0)}$$

$$= \frac{\pi x}{2} \left(1 - \frac{x^2}{2^2} \right) \frac{I_2(x)}{I_2(0)}$$

$$= \frac{\pi x}{2} \left(1 - \frac{x^2}{2^2} \right) \left(1 - \frac{x^2}{4^2} \right) \frac{I_4(x)}{I_4(0)}$$

$$= \frac{\pi x}{2} P_n(x) Q_{2n}(x),$$

where

$$P_n(x) = \prod_{j=1}^{n} \left(1 - \frac{x^2}{(2j)^2} \right)$$

and

$$Q_n(x) = \frac{I_n(x)}{I_n(0)}$$

for any $n \in \mathbb{N}$. Similarly, Eqs. (2.6.5) and (2.6.6) give

$$\cos \frac{\pi x}{2} = (1 - x^2) \frac{I_1(x)}{I_1(0)}$$

$$= (1 - x^2) \left(1 - \frac{x^2}{3^2} \right) \frac{I_3(x)}{I_3(0)}$$

$$= C_n(x) Q_{2n-1}(x),$$

where

$$C_n(x) = \prod_{j=1}^{n} \left(1 - \frac{x^2}{(2j-1)^2} \right).$$

For any $x \in \mathbb{R}$, the series $\sum_{j=1}^{\infty} x^2/j^2$ converges and hence the series

$$\sum_{j=1}^{\infty} \frac{x^2}{(2j)^2}$$

and

$$\sum_{j=1}^{\infty} \frac{x^2}{(2j-1)^2}$$

converge. Corollary 2.2.3 thus implies that the infinite products

$$P(x) = \prod_{j=1}^{\infty} \left(1 - \frac{x^2}{4j^2} \right)$$

and

$$C(x) = \prod_{j=1}^{\infty} \left(1 - \frac{x^2}{(2j-1)^2} \right)$$

converge for all $x \in \mathbb{R}$.

We now show that $Q_n(x) \to 1$ as $n \to \infty$. Consider the difference

$$I_n(0) - I_n(x) = \int_0^{\pi/2} (1 - \cos x\xi) \cos^n \xi \, d\xi > 0,$$

where $n > 0$. From the Maclaurin series for $\cos x\xi$ we know that

$$1 - \cos x\xi \leq \frac{1}{2} x^2 \xi^2;$$

therefore,

$$I_n(0) - I_n(x) \leq \frac{x^2}{2} \int_0^{\pi/2} \xi^2 \cos^n \xi \, d\xi. \tag{2.6.7}$$

Next, note that the function $\sin \xi - \xi \cos \xi$ is increasing on the interval $[0, \pi/2]$, as its derivative is the function $\xi \sin \xi$. Consequently

$$\sin \xi \geq \xi \cos \xi$$

for each $\xi \in [0, \pi/2]$. Inequality (2.6.7) therefore implies that

$$
\begin{aligned}
I_n(0) - I_n(x) &\leq \frac{x^2}{2} \int_0^{\pi/2} \xi \cos^{n-1} \xi \sin \xi \, d\xi \\
&= \frac{x^2}{2n} \int_0^{\pi/2} \xi n \cos^{n-1} \xi \sin \xi \, d\xi \\
&= \frac{x^2}{2n} \left(-\xi \cos^n \xi \Big|_0^{\pi/2} + \int_0^{\pi/2} \cos^n \xi \, d\xi \right) \\
&= \frac{x^2}{2n} I_n(0).
\end{aligned}
$$

We thus have

$$0 \le 1 - \frac{I_n(x)}{I_n(0)} \le \frac{x^2}{2n}.$$

For any $x \in \mathbb{R}$,

$$\lim_{n \to \infty} \frac{x^2}{2n} = 0,$$

and therefore $Q_n(x) \to 1$ as $n \to \infty$ for all $x \in \mathbb{R}$. All the subsequences of $\{Q_n(x)\}$ must converge to 1; therefore, for all $x \in \mathbb{R}$,

$$\begin{aligned}
\sin \frac{\pi x}{2} &= \lim_{n \to \infty} \sin \frac{\pi x}{2} \\
&= \lim_{n \to \infty} \frac{\pi x}{2} P_n(x) Q_{2n}(x) \\
&= \frac{\pi x}{2} P(x).
\end{aligned}$$

Similarly, for all $x \in \mathbb{R}$,

$$\cos \frac{\pi x}{2} = C(x).$$

Equations (2.6.1) and (2.6.2) follow from these expressions by replacing x with $2x/\pi$.   □

Example 2.6.1 The Wallis product is a simple consequence of Eq. (2.6.1). We have

$$1 = \sin \frac{\pi}{2} = \frac{\pi}{2} \prod_{j=1}^{\infty} \left(1 - \frac{\left(\frac{\pi}{2}\right)^2}{j^2 \pi^2} \right);$$

hence,

$$\begin{aligned}
\frac{2}{\pi} &= \prod_{j=1}^{\infty} \left(1 - \frac{1}{(2j)^2} \right) \\
&= \prod_{j=1}^{\infty} \frac{(2j)^2 - 1}{(2j)^2} \\
&= \prod_{j=1}^{\infty} \frac{(2j-1)(2j+1)}{(2j)^2}.
\end{aligned}$$

This equation implies

$$\frac{\pi}{2} = \frac{2}{1} \cdot \frac{2}{3} \cdot \frac{4}{3} \cdot \frac{4}{5} \cdot \frac{6}{5} \cdot \frac{6}{7} \cdots ,$$ (2.6.8)

which is the Wallis product.   △

Example 2.6.2 We use the Maclaurin series for $\sin x$ and $\log(1-x)$ along with the infinite product expansion for $\sin x$ to deduce the equation

$$\sum_{j=1}^{\infty} \frac{1}{j^2} = \frac{\pi^2}{6}.$$ (2.6.9)

The series for $\sin x$ gives

$$\frac{\sin x}{x} = \sum_{j=0}^{\infty} \frac{(-1)^j}{(2j+1)!} x^{2j}$$

$$= 1 - \frac{x^2}{6} + x^2 g(x)$$

for all $x \neq 0$, where

$$g(x) = \sum_{j=2}^{\infty} \frac{(-1)^j}{(2j+1)!} x^{2j-2}$$

for all x. Note that $g(0) = 0$. Moreover

$$0 < \frac{\sin x}{x} < 1$$

for all $x \in (-\pi, \pi) - \{0\}$. Letting

$$w(x) = 1 - \frac{\sin x}{x} = x^2 \left(\frac{1}{6} - g(x) \right)$$

for all $x \neq 0$, we find that if $0 < |x| < \pi$ then $0 < w(x) < 1$ and so

$$\frac{1}{x^2} \log \frac{\sin x}{x} = \frac{1}{x^2} \log(1 - w(x))$$

$$= -\frac{1}{x^2} \sum_{j=1}^{\infty} \frac{w^j(x)}{j}$$

$$= -\sum_{j=1}^{\infty} \frac{1}{j} x^{2j-2} \left(\frac{1}{6} - g(x) \right)^j$$

$$= -\left(\frac{1}{6} - g(x)\right) - \sum_{j=2}^{\infty} \frac{1}{j} x^{2j-2} \left(\frac{1}{6} - g(x)\right)^{j}$$

$$\rightarrow -\frac{1}{6} \qquad\qquad\qquad\qquad (2.6.10)$$

as $x \rightarrow 0$.

Let $0 < |x| < \pi$. Then

$$0 < \frac{x^2}{j^2 \pi^2} < 1$$

for all $j \geq 1$ and Eq. (2.6.1) implies that

$$\log \frac{\sin x}{x} = \sum_{j=1}^{\infty} \log\left(1 - \frac{x^2}{j^2 \pi^2}\right). \qquad\qquad (2.6.11)$$

Notice that this series is absolutely convergent, for it is evidently convergent and its terms are all negative (since $0 < |x| < \pi$). From Eq. (2.6.11) and the Maclaurin series for

$$\log\left(1 - \frac{x^2}{j^2 \pi^2}\right)$$

we have

$$\frac{1}{x^2} \log \frac{\sin x}{x} = -\sum_{j=1}^{\infty} \sum_{k=1}^{\infty} \frac{x^{2k-2}}{k j^{2k} \pi^{2k}},$$

and the absolute convergence of the series in Eq. (2.6.11) therefore implies that

$$\frac{1}{x^2} \log \frac{\sin x}{x} = \sum_{k=1}^{\infty} g_k(x)$$

where

$$g_k(x) = -\frac{x^{2k-2}}{k \pi^{2k}} \sum_{j=1}^{\infty} \frac{1}{j^{2k}}$$

for all $k \in \mathbb{N}$ and all x for which $0 < |x| < \pi$. It is plain that $\{g_n\}$ is a sequence of functions continuous at all x. Since

$$|g_k(x)| = \frac{1}{k\pi^2} \left(\frac{x}{\pi}\right)^{2k-2} \sum_{j=1}^{\infty} \frac{1}{j^{2k}}$$

and $\sum_{j=1}^{\infty} 1/j^{2k}$ converges for each k, it follows that $\sum_{k=1}^{\infty} g_k(x)$ is a power series with radius of convergence π. Therefore it is uniformly convergent to a continuous function (by Corollary 1.4.10) on any interval $[-\alpha, \alpha]$, where $0 < \alpha < \pi$. Thus

$$-\frac{1}{6} = \lim_{x \to 0} \frac{1}{x^2} \log \frac{\sin x}{x}$$

$$= \lim_{x \to 0} \left(-\frac{1}{\pi^2} \sum_{j=1}^{\infty} \frac{1}{j^2} + \sum_{k=2}^{\infty} g_k(x)\right)$$

$$= -\frac{1}{\pi^2} \sum_{j=1}^{\infty} \frac{1}{j^2}.$$

Equation (2.6.9) follows immediately.   △

Example 2.6.3 The Riemann zeta function ζ is given by

$$\zeta(z) = \sum_{j=1}^{\infty} \frac{1}{j^z}$$

for all $z \in \mathbb{C}$ such that Re $(z) > 1$. Example 2.6.2 shows that

$$\zeta(2) = \frac{\pi^2}{6}. \tag{2.6.12}$$

The same approach can be used to find $\zeta(2m)$ for any $m \in \mathbb{N}$. We know from Example 2.6.2 that

$$\log \frac{\sin x}{x} = -\sum_{j=1}^{\infty} \sum_{k=1}^{\infty} \frac{x^{2k}}{kj^{2k}\pi^{2k}}$$

whenever $0 < |x| < \pi$, and since the order of summation can be changed this equation gives

$$\log \frac{\sin x}{x} = -\sum_{k=1}^{\infty} \frac{\zeta(2k)}{k\pi^{2k}} x^{2k}. \tag{2.6.13}$$

Suppose that we wish to find the value of $\zeta(4)$. Equation (2.6.13) yields

$$\frac{1}{x^4}\left(\log\frac{\sin x}{x} + \frac{\zeta(2)}{\pi^2}x^2\right) = -\frac{\zeta(4)}{2\pi^4} - \sum_{k=3}^{\infty}\frac{\zeta(2k)}{k\pi^{2k}}x^{2k-4},$$

and therefore

$$-2\pi^4 \lim_{x\to 0}\frac{1}{x^4}\left(\log\frac{\sin x}{x} + \frac{x^2}{6}\right) = \zeta(4),$$

since $\zeta(2) = \pi^2/6$.

We must now evaluate the limit. First, recall that $\lim_{x\to 0}\sin x/x = 1$ and that $\log 1 = 0$. In the following calculation, each equation except the second and last is obtained by applying L'Hôpital's rule and simplifying the result:

$$\lim_{x\to 0}\frac{1}{x^4}\left(\log\frac{\sin x}{x} + \frac{x^2}{6}\right) = \lim_{x\to 0}\frac{\frac{x\cos x - \sin x}{x\sin x} + \frac{x}{3}}{4x^3}$$

$$= \frac{1}{12}\lim_{x\to 0}\frac{3(x\cos x - \sin x) + x^2\sin x}{x^4\sin x}$$

$$= \frac{1}{12}\lim_{x\to 0}\frac{x\cos x - \sin x}{4x^2\sin x + x^3\cos x}$$

$$= \frac{1}{12}\lim_{x\to 0}\frac{-\sin x}{8\sin x + 7x\cos x - x^2\sin x}$$

$$= \frac{1}{12}\lim_{x\to 0}\frac{-\cos x}{15\cos x - 9x\sin x - x^2\cos x}$$

$$= -\frac{1}{180}.$$

Hence

$$\zeta(4) = \frac{\pi^4}{90}. \tag{2.6.14}$$

The values for $\zeta(6), \zeta(8), \ldots, \zeta(2m)$ can be found in the same way, although the calculations of the limits become more cumbersome as m increases. An alternative approach is to note that

$$\frac{d}{dx}\log\frac{\sin x}{x} = \frac{\cos x}{\sin x} - \frac{1}{x}$$

$$= \cot x - \frac{1}{x}.$$

We therefore deduce from Eq. (2.6.13) that

$$x \cot x = 1 - \sum_{k=1}^{\infty} \frac{2\zeta(2k)}{\pi^{2k}} x^{2k}. \tag{2.6.15}$$

Thus we can find $\zeta(2k)$, for any $k \in \mathbb{N}$, from the Maclaurin series for $x \cot x$. It turns out that the coefficients for this Maclaurin series are related to Bernoulli numbers. Specifically,

$$\frac{x}{2} \cot \frac{x}{2} = 1 - \sum_{k=1}^{\infty} \frac{B_{2k} x^{2k}}{(2k)!}, \tag{2.6.16}$$

where each B_{2k} is a Bernoulli number. The first few even Bernoulli numbers are

$$B_2 = \frac{1}{6}, \quad B_4 = -\frac{1}{30}, \quad B_6 = \frac{1}{42}, \quad B_8 = -\frac{1}{30}, \quad B_{10} = \frac{5}{66}.$$

(For $k \geq 1$ we have $B_{2k+1} = 0$.) There are other methods for generating Bernoulli numbers (and different conventions for them). The sequence $\{B_n\}$ has been tabulated for a large number of values of n. Equations (2.6.15) and (2.6.16) imply

$$\zeta(2k) = \frac{(-1)^{k+1} 2^{2k-1} \pi^{2k} B_{2k}}{(2k)!}. \tag{2.6.17}$$

It is clear that this approach will not work for evaluating $\zeta(2k + 1)$, where $k \in \mathbb{N}$. In fact, there is no simple method or formula known for evaluating the Riemann zeta function at odd positive integers. $\triangle$

Exercises 2.6

1. a. Show that the product (2.6.1) converges uniformly on intervals of the form $[-\alpha, \alpha]$, where $0 < \alpha < \pi$. Show also that it converges uniformly on any interval $I = [a, b]$ that does not contain a multiple of π.
   b. Show that the product (2.6.2) converges uniformly on intervals I that do not contain points of the form $(2m - 1)\pi/2$, where m is an integer.
2. Given relation (2.6.1) and the double angle relation $\sin 2x = 2 \sin x \cos x$, derive relation (2.6.2).
3. Derive the relation

$$\cot x = \frac{1}{x} - \sum_{j=1}^{\infty} \frac{2x}{j^2 \pi^2 - x^2}, \tag{2.6.18}$$

for all $x \in [a, b]$, where $0 < a < b < \pi$. (This relation can obviously be extended to intervals that do not contain multiples of π.)
4. Use Eq. (2.6.18) to show that

$$\pi^2 \csc^2 \pi x = \sum_{j=0}^{\infty} \frac{1}{(x+j)^2} + \sum_{j=1}^{\infty} \frac{1}{(x-j)^2}, \qquad (2.6.19)$$

whenever $0 < x < 1$. (This equation can be extended to intervals that do not contain integers.)

2.7 Abel's Limit Theorem for Infinite Products

If the power series

$$f(x) = \sum_{j=0}^{\infty} a_j x^j$$

has a unit radius of convergence and the series $\sum_{j=0}^{\infty} a_j$ converges, then Abel's limit theorem for series (Theorem 1.5.8) shows that

$$\lim_{x \to 1^-} f(x) = \sum_{j=0}^{\infty} a_j.$$

If $\sum_{j=0}^{\infty} |a_j|$ converges, it follows from the Weierstrass M-test and Theorem 1.4.16 that $\sum_{j=0}^{\infty} a_j x^j$ is uniformly convergent on $[0, 1]$, as $|a_j x^j| \le |a_j|$. If $\sum_{j=0}^{\infty} a_j$ is conditionally convergent, then the modified Dirichlet test (Theorem 1.4.25) can be used to prove that the power series converges uniformly on $[0, 1]$, since

$$\sum_{j=0}^{\infty} |x^{j+1} - x^j| = 0$$

if $x = 1$ and

$$\sum_{j=0}^{\infty} |x^{j+1} - x^j| = \sum_{j=0}^{\infty} |x^j||x - 1| = (1 - x) \sum_{j=0}^{\infty} x^j = \frac{1-x}{1-x} = 1$$

if $0 \le x < 1$.

This result is not true for complex power series. For example, for any $n \in \mathbb{N}$ the power series

$$\sum_{j=1}^{\infty} \frac{z^{2^n} - z^{2^{n+1}}}{j}$$

converges to 0 at 1 but is unbounded in any neighbourhood of $e^{i\pi/2^n}$. Hence its value at 1 is not the limit as z approaches 1 in $D(0; 1)$.

In this section we consider the relationship between products of the form

$$P(x) = \prod_{j=1}^{\infty} (1 + a_j x) \tag{2.7.1}$$

and

$$Q(x) = \prod_{j=1}^{\infty} (1 + a_j x^j), \tag{2.7.2}$$

where $\{a_n\}$ is a sequence of numbers, and the product

$$\prod_{j=1}^{\infty} (1 + a_j).$$

Specifically, we look at conditions on $\{a_n\}$ which ensure that

$$\lim_{x \to 1^-} P(x) = \prod_{j=1}^{\infty} (1 + a_j) \tag{2.7.3}$$

and

$$\lim_{x \to 1^-} Q(x) = \prod_{j=1}^{\infty} (1 + a_j). \tag{2.7.4}$$

If the product

$$R(x) = \prod_{j=1}^{\infty} \left(1 + f_j(x)\right)$$

is uniformly convergent on an interval I throughout which $f_j(x) > -1$ for all j, and f_j is continuous on I for all j, then Theorem 2.5.2(1) implies that R is continuous on I. Then

$$\lim_{x \to c} R(x) = \prod_{j=1}^{\infty} \left(1 + f_j(c)\right)$$

for all $c \in I$. We thus look for conditions on $\{a_n\}$ which ensure that the products converge uniformly on an interval that contains 1. First we investigate the case where $\prod_{j=1}^{\infty}(1 + a_j)$ is regularly convergent. We assume throughout that $a_j > -1$.

Suppose therefore that the series $\sum_{j=0}^{\infty} a_j$ is absolutely convergent. Then it is plain that $|a_j x| \le |a_j|$ and $|a_j x^j| \le |a_j|$ for all $x \in [0, 1]$ and $j \in \mathbb{N} \cup \{0\}$; consequently, the series $\sum_{j=0}^{\infty} |a_j x|$ and $\sum_{j=0}^{\infty} |a_j x^j|$ converge uniformly on $[0, 1]$ to bounded functions. Moreover $a_j x > -1$ and $a_j x^j > -1$ since $a_j > -1$ and $0 \le x \le 1$. Theorems 2.5.1 and 2.5.2 thus imply that both P and Q are continuous on $[0, 1]$ and therefore Eqs. (2.7.3) and (2.7.4) are satisfied.

Suppose on the other hand that the series $\sum_{j=0}^{\infty} a_j$ is conditionally convergent and that $\sum_{j=0}^{\infty} a_j^2$ converges. Then

$$\sum_{j=0}^{\infty} a_j x = x \sum_{j=0}^{\infty} a_j,$$

and this series is uniformly convergent on $[0, 1]$ to a bounded function. Similarly $\sum_{j=0}^{\infty} a_j^2 x^2$ is uniformly convergent on $[0, 1]$ to a bounded function. Hence the product (2.7.1) converges uniformly on $[0, 1]$ by Cauchy's test (Theorem 2.5.4) and therefore we know that Eq. (2.7.3) holds.

Since $a_j \to 0$ as $j \to \infty$ there is a number Λ such that $|a_j| \le \Lambda$ for all j. Thus the series $\sum_{j=0}^{\infty} a_j x^j$ converges whenever $|x| < 1$. The conditional convergence of $\sum_{j=0}^{\infty} a_j$ implies that the radius of convergence of this power series must be 1. Abel's limit theorem (Theorem 1.5.8) shows that $\sum_{j=0}^{\infty} a_j x^j$ is uniformly convergent on $[0, 1]$, and it converges to a function that is continuous by Corollary 1.4.18, and hence bounded, on $[0, 1]$. Since $a_j^2 x^{2j} \le a_j^2$ whenever $|x| \le 1$, the Weierstrass M-Test implies that $\sum_{j=0}^{\infty} a_j^2 x^{2j}$ converges uniformly on $[0, 1]$ to a bounded function. The conditions of Theorem 2.5.4 are once again satisfied; hence Q is continuous on $[0, 1]$ and Eq. (2.7.4) holds.

In summary, we have the following result.

Theorem 2.7.1 *If the product $\prod_{j=1}^{\infty}(1 + a_j)$ is regularly convergent and $a_j > -1$ for all $j \in \mathbb{N}$, then the products $P(x)$ and $Q(x)$ are uniformly convergent on $[0, 1]$ and Eqs. (2.7.3) and (2.7.4) are satisfied.*

If the product $\prod_{j=1}^{\infty}(1 + a_j)$ is irregularly convergent, then Eqs. (2.7.3) and (2.7.4) are, in general, not satisfied. Hardy [29] gives two examples that offer a glimpse of some of the problems with irregularly convergent products.

Theorem 2.7.1 shows that if $\prod_{j=1}^{\infty}(1 + a_j)$ is regularly convergent and $a_j > -1$ for all $j \in \mathbb{N}$, then $P(x)$ must be uniformly convergent on $[0, 1]$. The next example shows that it can diverge for all $x \in (0, 1)$ if $\prod_{j=1}^{\infty}(1 + a_j)$ is irregularly convergent.

Example 2.7.1 Consider the product $\prod_{j=1}^{\infty}(1 + a_j)$, where

$$a_{2j} = \frac{1}{\sqrt{j} - \frac{1}{2}}, \quad a_{2j+1} = -\frac{1}{\sqrt{j} + \frac{1}{2}}.$$

We have

$$(1 + a_{2j})(1 + a_{2j+1}) = 1,$$

and therefore

$$\prod_{j=1}^{\infty}(1 + a_j) = 1.$$

This product is irregularly convergent (see Exercises 2.3, question 8). Suppose that we form the product $P(x)$ using this sequence $\{a_n\}$. Then

$$(1 + a_{2j}x)(1 + a_{2j+1}x) = 1 - \frac{x(x - 1)}{j - \frac{1}{4}}.$$

If the product $P(x)$ were to converge, then

$$\prod_{j=1}^{\infty}\left(1 - \frac{x(x - 1)}{j - \frac{1}{4}}\right) \tag{2.7.5}$$

would converge. The series

$$\sum_{j=1}^{\infty} \frac{x(x - 1)}{j - \frac{1}{4}},$$

however, diverges unless $x = 0$ or $x = 1$. We can apply Corollary 2.2.3 to show that the product (2.7.5) diverges for all $x \in (0, 1)$. The product $P(x)$ thus converges only for $x = 0$ and $x = 1$.   △

Hardy gives the following striking example of a product of the form (2.7.2) such that

$$\lim_{x \to 1^-} Q(x) = 2 \prod_{j=1}^{\infty}(1 + a_j). \tag{2.7.6}$$

Example 2.7.2 Let

$$a_j = \frac{e^{ij\phi}}{\sqrt{j}}$$

and

$$b_j = \frac{2 \cos j\phi}{\sqrt{j}} + \frac{1}{j}$$

for all $j \in \mathbb{N}$, where $\phi \in \mathbb{R}$ and ϕ is not a multiple of π. Thus

$$a_1 = \cos \phi + i \sin \phi \neq -1;$$

hence $a_j \neq -1$ for all $j \in \mathbb{N}$. Moreover, for all $j \in \mathbb{N}$ and $x \in [0, 1)$ we have

$$|a_j x^j| = \frac{x^j}{\sqrt{j}} < 1.$$

Thus $a_j x^j \neq -1$, so that

$$|1 + a_j x^j| > 0.$$

This inequality therefore holds for all $x \in [0, 1]$ and $j \in \mathbb{N}$.

Example 2.3.9 shows that the infinite product

$$\prod_{j=1}^{\infty} (1 + a_j)$$

converges regularly, but that the product

$$\prod_{j=1}^{\infty} (1 + b_j)$$

converges irregularly. The products are related by

$$\prod_{j=1}^{\infty} |1 + a_j|^2 = \prod_{j=1}^{\infty} (1 + b_j).$$

It can be shown that $\prod_{j=1}^{\infty} (1 + a_j x^j)$ converges for all $x \in [0, 1]$ and

$$\lim_{x \to 1-} \prod_{j=1}^{\infty} \left(1 + a_j x^j\right) = \prod_{j=1}^{\infty} (1 + a_j). \qquad (2.7.7)$$

(See Hardy (*op. cit.*) for the full details.) It follows that the product

$$\prod_{j=1}^{\infty} |1 + a_j x^j|^2 = \prod_{j=1}^{\infty} (1 + a_j x^j)\overline{1 + a_j x^j}$$

converges for all $x \in [0, 1]$ (to nonzero numbers), since

$$\overline{1 + a_j x^j} = 1 + \overline{a}_j x^j = 1 + \frac{e^{-ij\phi} x^j}{\sqrt{j}},$$

and

$$\lim_{x \to 1^-} \prod_{j=1}^{\infty} |1 + a_j x^j|^2 = \lim_{x \to 1^-} \prod_{j=1}^{\infty} (1 + a_j x^j) \prod_{j=1}^{\infty} \overline{1 + a_j x^j}$$

$$= \prod_{j=1}^{\infty} (1 + a_j) \prod_{j=1}^{\infty} \overline{1 + a_j}$$

$$= \prod_{j=1}^{\infty} |1 + a_j|^2.$$

For all $j \in \mathbb{N}$ and $x \in [0, 1]$ we have

$$0 < |1 + a_j x^j|^2$$

$$= \left| 1 + \frac{e^{ij\phi} x^j}{\sqrt{j}} \right|^2$$

$$= \left(1 + \frac{x^j \cos j\phi}{\sqrt{j}} \right)^2 + \frac{x^{2j} \sin^2 j\phi}{j}$$

$$= 1 + A_j(x),$$

where

$$A_j(x) = \frac{2x^j \cos j\phi}{\sqrt{j}} + \frac{x^{2j}}{j},$$

so that $A_j(x) > -1$ and

$$\lim_{x \to 1^-} \prod_{j=1}^{\infty} (1 + A_j(x)) = \prod_{j=1}^{\infty} |1 + a_j|^2 = \prod_{j=1}^{\infty} (1 + b_j). \qquad (2.7.8)$$

Let $B_n(x) = b_n x^n$ for each $n \in \mathbb{N}$ and $x \in [0, 1]$. The sequence $\{b_n\}$ is evidently bounded, and so $\sum_{j=1}^{\infty} |B_j(x)|$ converges for all $x \in [0, 1)$. Moreover, as

$$1 + b_n = |1 + a_n|^2 > 0$$

for each $n \in \mathbb{N}$, we have $b_n > -1$, so that $B_n(x) > -1$ for each $n \in \mathbb{N}$ and each $x \in [0, 1)$. Corollary 2.2.7 therefore shows that the product

$$\prod_{j=1}^{\infty} (1 + B_j(x))$$

converges for all $x \in [0, 1)$. Define the function R by

$$R(x) = \prod_{j=1}^{\infty} \frac{1 + B_j(x)}{1 + A_j(x)}$$

for all $x \in [0, 1)$. This product must converge since $\prod_{j=1}^{\infty} (1 + A_j(x))$ converges to a function that is nonzero throughout the interval $[0, 1)$. It will be shown that

$$\lim_{x \to 1^-} R(x) = 2.$$

For each $j \in \mathbb{N}$ and $x \in [0, 1]$ we have

$$\frac{1 + B_j(x)}{1 + A_j(x)} = \frac{j + 2x^j \sqrt{j} \cos j\phi + x^j}{j + 2x^j \sqrt{j} \cos j\phi + x^{2j}}$$

$$= 1 + \frac{x^j - x^{2j}}{j + 2x^j \sqrt{j} \cos j\phi + x^{2j}}.$$

Therefore the function R can be written as

$$R(x) = \prod_{j=1}^{\infty} (1 + C_j(x)),$$

where

$$C_j(x) = \frac{x^j (1 - x^j)}{j + 2x^j \sqrt{j} \cos j\phi + x^{2j}}$$

for all $j \in \mathbb{N}$ and $x \in [0, 1]$. Note that

$$j + 2x^j \sqrt{j} \cos j\phi + x^{2j} = j(1 + A_j(x)) > 0,$$

so that $C_j(x) > 0$ for all $x \in (0, 1)$ and $j \in \mathbb{N}$, and

$$jC_j(x) = \frac{x^j(1-x^j)}{1+A_j(x)}. \tag{2.7.9}$$

Now the function $1 + A_1(x)$ is positive and continuous on the closed interval $[0, 1]$, and so there exists $M > 0$ such that $1 + A_1(x) > M$ for all $x \in [0, 1]$. For each $j > 1$, we have

$$
\begin{aligned}
1 + A_j(x) &= 1 + \frac{2x^j \cos j\phi}{\sqrt{j}} + \frac{x^{2j}}{j} \\
&\geq 1 - \frac{2x^j}{\sqrt{j}} + \frac{x^{2j}}{j} \\
&= \left(1 - \frac{x^j}{\sqrt{j}}\right)^2 \\
&\geq \left(1 - \frac{1}{\sqrt{2}}\right)^2 .
\end{aligned}
$$

As $x^j(1-x^j) < 1$ for each $x \in [0, 1]$, we therefore conclude that there is a number M_1 such that

$$jC_j(x) < M_1$$

for all $x \in [0, 1]$ and $j \in \mathbb{N}$. Since $C_j(x) < M_1/j$ for all such x and j, the series $\sum_{j=1}^{\infty} C_j^2(x)$ is uniformly convergent on $[0, 1]$ by the Weierstrass M-test. Moreover $C_j(x) \leq 1/2$ for all $x \in [0, 1]$ and $j \geq 2M_1$, and so the argument in the proof of Theorem 2.3.1 can be used to show that

$$|\log(1 + C_j(x)) - C_j(x)| \leq C_j^2(x)$$

for all such x and j; hence the series

$$\sum_{j=1}^{\infty} |\log(1 + C_j(x)) - C_j(x)|$$

is uniformly convergent on $[0, 1]$ by the comparison test. Therefore Corollary 1.4.18 shows that

$$\lim_{x \to 1^-} \sum_{j=1}^{\infty} |\log(1 + C_j(x)) - C_j(x)| = \sum_{j=1}^{\infty} |\log(1 + C_j(1)) - C_j(1)| = 0,$$

and so

$$\lim_{x \to 1^-} \sum_{j=1}^{\infty} (\log(1 + C_j(x)) - C_j(x)) = 0. \tag{2.7.10}$$

As the product $R(x)$ converges for all $x \in [0, 1)$ and $C_j(1) = 0$ for all $j \in \mathbb{N}$, the series $\sum_{j=1}^{\infty} C_j(x)$ converges for all $x \in [0, 1]$ by Theorem 2.2.1, though the function it represents might not be continuous at 1. Let

$$D_j(x) = \frac{x^j(1 - x^j)}{j}$$

for each $j \in \mathbb{N}$ and $x \in [0, 1]$. Then

$$
\begin{aligned}
D_j(x) - C_j(x) &= \frac{x^j(1 - x^j)}{j} - \frac{x^j(1 - x^j)}{j + 2x^j\sqrt{j}\cos j\phi + x^{2j}} \\
&= \frac{x^j(1 - x^j)}{j} \cdot \frac{2x^j\sqrt{j}\cos j\phi + x^{2j}}{j + 2x^j\sqrt{j}\cos j\phi + x^{2j}}.
\end{aligned}
$$

As

$$
\begin{aligned}
j^{\frac{3}{2}}|D_j(x) - C_j(x)| &= \frac{x^j(1 - x^j)\left|2x^j\cos j\phi + \frac{x^{2j}}{\sqrt{j}}\right|}{1 + A_j(x)} \\
&= jC_j(x)\left|2x^j\cos j\phi + \frac{x^{2j}}{\sqrt{j}}\right| \\
&< 3M_1,
\end{aligned}
$$

we have

$$|D_j(x) - C_j(x)| < \frac{3M_1}{j^{3/2}}$$

for all $x \in [0, 1]$ and $j \in \mathbb{N}$. This inequality shows that the series

$$\sum_{j=1}^{\infty} (D_j(x) - C_j(x))$$

converges uniformly on $[0, 1]$, by the Weierstrass M-test, to a function that must be continuous on $[0, 1]$. Consequently

$$\lim_{x \to 1^-} \sum_{j=1}^{\infty} (D_j(x) - C_j(x)) = 0.$$

As $\sum_{j=1}^{\infty} C_j(x)$ also converges for all $x \in [0, 1]$, so does the series $\sum_{j=1}^{\infty} D_j(x)$, and therefore

$$\lim_{x \to 1^-} \left(\sum_{j=1}^{\infty} D_j(x) - \sum_{j=1}^{\infty} C_j(x) \right) = 0. \tag{2.7.11}$$

For all x such that $|x| < 1$,

$$\sum_{j=1}^{\infty} \frac{x^j}{j} = -\log(1 - x)$$

and

$$\sum_{j=1}^{\infty} \frac{x^{2j}}{j} = -\log(1 - x^2);$$

consequently,

$$\sum_{j=1}^{\infty} D_j(x) = -\log(1 - x) + \log(1 - x^2)$$

$$= \log \frac{1 - x^2}{1 - x}$$

$$= \log(1 + x)$$

$$= \sum_{j=1}^{\infty} \frac{(-1)^{j+1} x^j}{j},$$

and since the series $\sum_{j=1}^{\infty} (-1)^{j+1}/j$ converges, Abel's limit theorem gives

$$\lim_{x \to 1^-} \sum_{j=1}^{\infty} D_j(x) = \log 2.$$

Equation (2.7.11) implies that

$$\lim_{x \to 1^-} \sum_{j=1}^{\infty} C_j(x) = \log 2,$$

and therefore it follows from Eq. (2.7.10) that

$$\lim_{x \to 1^-} \sum_{j=1}^{\infty} \log(1 + C_j(x)) = \log 2.$$

This equation shows that

$$\lim_{x \to 1^-} R(x) = \lim_{x \to 1^-} \prod_{j=1}^{\infty} (1 + C_j(x)) = 2,$$

and therefore

$$\lim_{x \to 1^-} \prod_{j=1}^{\infty} (1 + b_j x^j) = \lim_{x \to 1^-} \prod_{j=1}^{\infty} (1 + B_j(x))$$

$$= 2 \lim_{x \to 1^-} \prod_{j=1}^{\infty} (1 + A_j(x))$$

$$= 2 \prod_{j=1}^{\infty} (1 + b_j)$$

by Eq. (2.7.8).   △

2.8   Weierstrass Products

Let $\{z_0, z_1, \ldots\}$ be a set S of complex numbers, where $z_0 = 0$, and let $\{m_0, m_1, \ldots\}$ be a set M of non-negative integers. We wish to construct an analytic function whose zeros are the members of S with corresponding multiplicities in M. Should S be finite of cardinality n, the product

$$\prod_{j=0}^{n-1} (z - z_j)^{m_j}$$

would be such a function, but the corresponding product might not converge if S is infinite. For instance, the product

$$\prod_{j=0}^{\infty} (z - j)^{m_j}$$

diverges since the sequence $\{z - n\}$ does not approach 1, for any fixed z.

An alternative is to consider the function

$$z^{m_0} \prod_{j=1}^{\infty} \left(1 - \frac{z}{z_j} \right)^{m_j},$$

which also has the same zeros with the required multiplicities. However this product might not converge either. Let us consider the following example.

Example 2.8.1 The product

$$\prod_{j=1}^{\infty} \left(1 - \frac{z}{j} \right)$$

has a zero at every positive integer. Corollary 2.2.3 shows that this product diverges for all $z \neq 0$. However we can introduce the "scaling factor" $e^{z/j}$ to give the product

$$\prod_{j=1}^{\infty} \left(1 - \frac{z}{j} \right) e^{z/j},$$

which has the same zeros. We now prove that this product converges.

Note first that if $|z| < j$ then

$$\log \left(\left(1 - \frac{z}{j} \right) e^{z/j} \right) = \log \left(1 - \frac{z}{j} \right) + \frac{z}{j}$$

$$= -\sum_{k=1}^{\infty} \frac{z^k}{k j^k} + \frac{z}{j}$$

$$= -\sum_{k=2}^{\infty} \frac{z^k}{k j^k}.$$

Thus the effect of the scaling factor is to cancel the term $-z/j$ in the series for the logarithm.

The proof of Theorem 2.3.1 shows that if $|z| \leq 1/2$ then

$$|\log(1 + z) - z| \leq |z|^2.$$

Thus if we choose R such that $|z| \leq R$ and $j \geq 2R$ then

$$\left| \log \left(\left(1 - \frac{z}{j} \right) e^{z/j} \right) \right| = \left| \log \left(1 - \frac{z}{j} \right) + \frac{z}{j} \right|$$

$$\leq \left| \frac{z}{j} \right|^2.$$

$$\leq \frac{R^2}{j^2}.$$

Therefore, if we denote by $\lceil R \rceil$ the least integer greater than or equal to R, then the Weierstrass M-test shows that the series

$$\sum_{j=2\lceil R \rceil}^{\infty} \left| \log\left(\left(1 - \frac{z}{j}\right) e^{z/j}\right)\right|$$

is uniformly convergent to a bounded function on the compact disc $\{z : |z| \leq R\}$ and hence on any compact subset of $\mathbb{C}$. By Corollary 2.4.3 the product

$$\prod_{j=1}^{\infty} \left(1 - \frac{z}{j}\right) e^{z/j}$$

therefore converges uniformly to an analytic function on any compact subset of $\mathbb{C}$. It converges to 0 at every $z \in \mathbb{N}$. $\triangle$

The idea of scaling factors was introduced by Weierstrass.

Definition 2.8.1 For each non-negative integer k and each $z \in \mathbb{C}$, define

$$E_k(z) = (1 - z)e^{S_k(z)},$$

where

$$S_k(z) = \sum_{j=1}^{k} \frac{z^j}{j}.$$

Then $E_k(z)$ is a **Weierstrass primary factor**.

Note that $S_0(z) = 0$, so that $E_0(z) = 1 - z$. Moreover, the first factor of the product considered in Example 2.8.1 is

$$(1 - z)e^z = E_1(z).$$

Observe also that E_k is an entire function with a zero of order 1 at 1 and no other zeros.

For all z such that $|z| < 1$ the power series $\sum_{j=1}^{\infty} z^j/j$ converges uniformly to $-\log(1 - z)$. Since

$$e^{-\log(1-z)} = \frac{1}{1 - z},$$

it follows that the sequence $\{E_k(z)\}$ converges uniformly to 1 on the disc $D(0; 1)$. Note also that $E_k(0) = 1$ for all k.

Lemma 2.8.1 *If* $|z| \leq 1$ *then*

$$|1 - E_k(z)| \leq |z|^{k+1}.$$

Proof As equality holds if $k = 0$ or $z \in \{0, 1\}$, we may assume that $k \geq 1$, $z \neq 0$ and $z \neq 1$. Differentiation gives

$$E_k'(z) = -e^{S_k(z)} + (1 - z)S_k'(z)e^{S_k(z)}$$

$$= e^{S_k(z)}\left(-1 + (1 - z)\sum_{j=0}^{k-1} z^j\right)$$

$$= e^{S_k(z)}\left(-1 + (1 - z)\frac{1 - z^k}{1 - z}\right)$$

$$= -z^k e^{S_k(z)},$$

so that

$$(1 - E_k(z))' = z^k e^{S_k(z)}. \tag{2.8.1}$$

Thus $(1 - E_k(z))'$ has a zero of order k at 0. Since $1 - E_k(0) = 0$, we conclude that $1 - E_k(z)$ has a zero of order $k + 1$ at 0. Therefore the function $(1 - E_k(z))/z^{k+1}$ has a Maclaurin series expansion $\sum_{j=0}^{\infty} c_j z^j$ at all $z \in \mathbb{C}$. Thus

$$\sum_{j=0}^{\infty} c_j z^j = \frac{1 - E_k(z)}{z^{k+1}}.$$

Equation (2.8.1) shows that the coefficients in the Maclaurin series of $(1 - E_k(z))'$ are non-negative real numbers, since

$$(1 - E_k(z))' = z^k \sum_{l=0}^{\infty} \frac{1}{l!}\left(\sum_{j=1}^{k} \frac{z^j}{j}\right)^l.$$

Consequently the same is true of c_j for all j. For all z such that $0 < |z| < 1$, we therefore obtain

$$\left|\frac{1 - E_k(z)}{z^{k+1}}\right| \leq \sum_{j=0}^{\infty} |c_j||z^j|$$

$$\leq \sum_{j=0}^{\infty} c_j$$

$$= 1 - E_k(1)$$

$$= 1,$$

and the result follows. $\square$

We also need the following easy lemma.

Lemma 2.8.2 *Suppose that $\{z_n\}$ is a sequence of nonzero complex numbers such that $\lim_{n\to\infty} |z_n| = \infty$. Then for any $r \geq 0$ the series $\sum_{j=1}^{\infty} r^j/|z_j|^j$ converges.*

Proof The hypothesis shows that for all $r \geq 0$ there exists N such that $|z_n| > 2r$ for all $n > N$. For all $j > N$ it follows that $r/|z_j| < 1/2$. The series $\sum_{j=1}^{\infty} (1/2)^j$ converges and the series in question consequently converges by the comparison test.
$\square$

Theorem 2.8.3 (Weierstrass) *Suppose that $\{z_n\}$ is a sequence of nonzero complex numbers such that $\lim_{n\to\infty} |z_n| = \infty$. Then there exists a sequence $\{k_j\}$ of non-negative integers such that the equation*

$$P(z) = \prod_{j=1}^{\infty} E_{k_j}\left(\frac{z}{z_j}\right)$$

defines an entire function. Moreover $P(z) = 0$ if and only if $z = z_n$ for some n.

Proof Choose $r > 0$ and any sequence $\{k_j\}$ of non-negative integers such that

$$\sum_{j=1}^{\infty} \left(\frac{r}{|z_j|}\right)^{k_j+1} \tag{2.8.2}$$

converges. For example, Lemma 2.8.2 shows that we may take $k_j = j - 1$ for all j. Now choose z such that $|z| \leq r$. There exists N such that $|z_j| > |z|$ for all $j \geq N$. For each $j \geq N$ we have

$$\left|1 - E_{k_j}\left(\frac{z}{z_j}\right)\right| \leq \left|\frac{z}{z_j}\right|^{k_j+1} \leq \left|\frac{r}{z_j}\right|^{k_j+1}.$$

The convergence of the series (2.8.2) therefore implies the uniform convergence of

$$\sum_{j=N}^{\infty} \left|1 - E_{k_j}\left(\frac{z}{z_j}\right)\right|$$

on $D(0; r)$ to a bounded function, by the Weierstrass M-test. Since r is arbitrary, the latter series is therefore uniformly convergent on every compact subset of $\mathbb{C}$. It consequently follows from Corollary 2.4.5 that the function P is entire. The remaining assertion is immediate from the fact that 1 is the only zero of E_{k_j} for each j.   □

Remark 1 The function P of Weierstrass's theorem is entire with the prescribed nonzero zeros and their prescribed multiplicities, because if m terms of the sequence $\{z_n\}$ are equal to a number a then a is a zero of P of multiplicity m. To obtain an entire function with 0 as a zero of multiplicity m_0, we simply multiply $P(z)$ by z^{m_0}. Thus if

$$f(z) = z^{m_0} \prod_{j=1}^{\infty} E_{k_j}\left(\frac{z}{z_j}\right)$$

for all z, then f is an entire function with the prescribed zeros and their multiplicities. In fact, so is $e^{g(z)} f(z)$ for any entire function g. For the sequence $\{j-1\}$ of non-negative integers we obtain

$$f(z) = z^{m_0} \prod_{j=1}^{\infty} E_{j-1}\left(\frac{z}{z_j}\right)$$

$$= z^{m_0} \prod_{j=1}^{\infty} \left(1 - \frac{z}{z_j}\right) \exp\left(\sum_{k=1}^{j-1} \frac{z^k}{k z_j^k}\right).$$

Remark 2 The sequence $\{k_j\}$ is chosen so that the series (2.8.2) converges for all $r > 0$. If the sequence $\{|z_j|\}$ grows sufficiently rapidly, we may be able to choose $\{k_j\}$ to be a constant sequence. For example, if $|z_j| = j$ for all j then we may take $k_j = 1$ for all j, since the series $\sum_{j=1}^{\infty} 1/j^2$ converges. The resulting product is

$$P(z) = \prod_{j=1}^{\infty} E_1\left(\frac{z}{z_j}\right)$$

$$= \prod_{j=1}^{\infty} \left(1 - \frac{z}{z_j}\right) e^{z/z_j}$$

for all z. If $\{k_j\}$ may be chosen to be a constant sequence, then let ρ be the least such constant. In this case the product is

$$\prod_{j=1}^{\infty} E_\rho\left(\frac{z}{z_j}\right)$$

for all z, and is called the **canonical product** of **rank** ρ. For instance, the rank of the product we obtained when $|z_j| = j$ for all j is 1.

Example 2.8.2 Let $z_j = \sqrt{j}$ for all j. Thus

$$\sum_{j=1}^{\infty} \frac{1}{|z_j|^{\rho+1}} = \sum_{j=1}^{\infty} \frac{1}{j^{(\rho+1)/2}}.$$

This series converges if and only if

$$\frac{\rho+1}{2} > 1.$$

The smallest integer ρ that satisfies this inequality is 2. Therefore we can find a canonical product of rank 2. It is

$$\prod_{j=1}^{\infty} E_2\left(\frac{z}{\sqrt{j}}\right) = \prod_{j=1}^{\infty}\left(1 - \frac{z}{\sqrt{j}}\right) \exp\left(\frac{z}{\sqrt{j}} + \frac{z^2}{2j}\right).$$

$$\triangle$$

On the other hand, it may be that the sequence $\{|z_j|\}$ grows too slowly for this idea to work. Such is the case if $|z_j| = \log j$ for all j, for instance, since $\sum_{j=2}^{\infty} 1/\log^m j$ diverges for any positive integer m.

Exercises 2.8

1. Find the canonical product representation for each of the following choices of z_j:

   (a) p^j, where $p > 1$;
   (b) j^p, where $p > 0$;
   (c) $j \log^2 j$.

2.9   The Weierstrass Factorization Theorem

We aim to show that every nonzero entire function can be written as a product involving Weierstrass primary factors.

Let $Z(f)$ be the set of zeros of a nonzero entire function f. Then $Z(f)$ has no accumulation point. This fact follows from the identity theorem, which asserts that if functions f and g are analytic on a connected open set D and $f = g$ on a subset of D that has an accumulation point, then $f = g$. Consequently $Z(f)$ is countable. We distinguish the following three cases:

1. $Z(f) = \emptyset$;

2. $Z(f)$ is non-empty but finite;
3. $Z(f) = \{z_1, z_2, \ldots\}$.

Theorem 2.9.1 *Let f be an entire function with no zeros. Then there exists an entire function g such that*

$$f(z) = e^{g(z)}$$

for all z.

Proof Since f has no zeros, the function f'/f is entire and therefore has an antiderivative F. Now

$$\left(f(z)e^{-F(z)}\right)' = f'(z)e^{-F(z)} - f(z)F'(z)e^{-F(z)}$$

$$= f'(z)e^{-F(z)} - f(z)\frac{f'(z)}{f(z)}e^{-F(z)}$$

$$= 0.$$

Thus $f(z)e^{-F(z)} = c$ for some constant c, so that

$$f(z) = ce^{F(z)}$$

for each z. Moreover, $c \neq 0$ since f is nonzero. As the exponential function is a surjection onto $\mathbb{C} - \{0\}$, we can write $c = e^b$ for some constant b. Hence

$$f(z) = e^{F(z)+b}$$

for every z, as required, since F is entire. $\qquad\qquad\qquad\qquad\qquad\qquad\Box$

We have now dealt with case (1). In case (3), which we will consider next, we assume that the elements of $Z(f)$ are listed in order of non-decreasing modulus. Since $Z(f)$ has no accumulation point, it then follows that

$$\lim_{n\to\infty} |z_n| = \infty.$$

The following theorem is known as the Weierstrass factorization theorem.

Theorem 2.9.2 *Let f be a nonzero entire function. Suppose that f has a zero of multiplicity m_0 at 0 and a sequence $(z_1, z_2, \ldots)$ of nonzero zeros in which each zero appears as many times as its multiplicity. There there exist an entire function g and a sequence $(k_1, k_2, \ldots)$ of non-negative integers such that*

$$f(z) = e^{g(z)}z^{m_0}\prod_{j=1}^{\infty} E_{k_j}\left(\frac{z}{z_j}\right)$$

for each z.

Proof It follows from Weierstrass's theorem that there is a sequence $\{k_j\}$ of non-negative integers such that

$$P(z) = z^{m_0} \prod_{j=1}^{\infty} E_{k_j}\left(\frac{z}{z_j}\right)$$

defines an entire function with the same zeros and multiplicities as f. Thus the function $Q = f/P$ may be extended to an entire function $\hat{Q}$ that has no zeros. By Theorem 2.9.1 there is an entire function g such that $\hat{Q}(z) = e^{g(z)}$ for all z. The result follows, since both sides of the equation are 0 on $\{0, z_1, z_2, \ldots\}$.   □

Case (2) is similarly disposed of by the following theorem. Its proof is analogous to that of the previous theorem and is left to the reader.

Theorem 2.9.3 *Let f be a nonzero entire function. Suppose that f has a zero of multiplicity m_0 at 0 and nonzero zeros $z_1, z_2, \ldots, z_n$, where each zero is listed as many times as its multiplicity. Then there exists an entire function g such that*

$$f(z) = e^{g(z)} z^{m_0} \prod_{j=1}^{n}\left(1 - \frac{z}{z_j}\right)$$

for each z.

Suppose that in the Weierstrass factorization of f the function g is a polynomial and the product

$$\prod_{j=1}^{\infty} E_{k_j}\left(\frac{z}{z_j}\right)$$

is canonical with rank ρ. Then we define the **Laguerre genus** $\varphi(f)$ of f as

$$\max\{\deg(g), \rho\}.$$

In the case where f has only finitely many zeros, we define $\varphi(f) = \deg(g)$. It can be shown [27, p. 216] that an entire function f such that $\varphi(f) \leq k$ can be written as

$$f(z) = e^{g(z)} z^{m_0} \prod_{j=1}^{\infty} E_k\left(\frac{z}{z_j}\right)$$

where g is a polynomial of degree at most k. Laguerre used the genus to obtain several results concerning the zeros of the derivatives of entire functions under certain conditions. For instance, let f be a non-constant entire function of genus 0 or 1, and suppose that $f(z)$ is real whenever z is real. If the zeros of f are all real, then so are the zeros of f'.

Example 2.9.1 If $f(z) = e^{z^n}$ for all z, then $\varphi(f) = n$.   △

Example 2.9.2 If f is a polynomial, then g is a constant function. In this case $\varphi(f) = 0$.   △

Example 2.9.3 The zeros of $\sin z$ are $\pm k\pi$, where $k \in \mathbb{N} \cup \{0\}$. For each positive integer n let us define $z_{2n-1} = n\pi$ and $z_{2n} = -n\pi$. Note that $\sum_{j=1}^{\infty} 1/|z_j|^2$ converges, and so we may write $\sin z$ in terms of a canonical product of rank 1. Thus

$$\sin z = e^{g(z)} z \prod_{j=1}^{\infty} E_1\left(\frac{z}{z_j}\right)$$

$$= e^{g(z)} z \prod_{j=1}^{\infty} \left(1 - \frac{z}{z_j}\right) e^{z/z_j}$$

$$= e^{g(z)} z \prod_{j=1}^{\infty} \left(1 - \frac{z}{j\pi}\right) e^{z/(j\pi)} \left(1 + \frac{z}{j\pi}\right) e^{-z/(j\pi)} \qquad (2.9.1)$$

$$= e^{g(z)} z \prod_{j=1}^{\infty} \left(1 - \frac{z^2}{j^2\pi^2}\right).$$

We claim that $e^{g(z)} = 1$. Let

$$P_n(z) = e^{g(z)} z \prod_{j=1}^{n} \left(1 - \frac{z^2}{j^2\pi^2}\right)$$

for all positive integers n and all z. Then $P_n(z)$ converges uniformly to $\sin z$ on any disc. Note also that $\{P_n'(z)\}$ converges to $\cos z$. Thus if z is not an integer multiple of π then $P_n'(z)/P_n(z)$ converges to $\cot z$. Now

$$\frac{P_n'(z)}{P_n(z)} = \frac{d}{dz} \log P_n(z)$$

$$= \frac{d}{dz} \left(g(z) + \log z + \sum_{j=1}^{n} \log\left(1 - \frac{z^2}{j^2\pi^2}\right)\right)$$

$$= g'(z) + \frac{1}{z} + \sum_{j=1}^{n} \frac{2z}{z^2 - j^2\pi^2}$$

$$\to g'(z) + \cot z$$

as $n \to \infty$ [39]. Hence $g'(z) = 0$, so that $g(z) = c$ for some constant c. Therefore

$$\frac{\sin z}{z} = e^c \prod_{j=1}^{\infty} \left(1 - \frac{z^2}{j^2 \pi^2} \right)$$

for all z such that $0 < |z| < \pi$. Taking limits as z approaches 0, we find that $e^c = 1$, as claimed. We conclude that

$$\sin z = z \prod_{j=1}^{\infty} \left(1 - \frac{z^2}{j^2 \pi^2} \right)$$

for all z, in agreement with Theorem 2.6.1, and that the Laguerre genus of this function is 1. We also observe from Eq. (2.9.1) that

$$\sin \pi z = \pi z \prod_{j=1}^{\infty} \left(1 + \frac{z}{j} \right) e^{-z/j} \left(1 - \frac{z}{j} \right) e^{z/j}$$

$$= \pi z \prod_{j=-\infty}^{-1} \left(1 - \frac{z}{j} \right) e^{z/j} \prod_{j=1}^{\infty} \left(1 - \frac{z}{j} \right) e^{z/j}.$$

$$\triangle$$

Recall that a meromorphic function is one whose singularities are isolated points which are poles of the function.

Theorem 2.9.4 *Every meromorphic function over $\mathbb{C}$ is a quotient of entire functions.*

Proof Let Q be a meromorphic function over $\mathbb{C}$. By Theorem 2.8.3 we can find an entire function g such that a is a pole of Q of order n if and only if a is a zero of g of order n. Thus the function gQ has only removable singularities and hence can be extended to an entire function f. Therefore $Q = f/g$.   □

Remark Theorems 2.8.3 and 2.9.2 can be generalized to functions defined on any open subset of $\mathbb{C}$ [10].

Exercises 2.9

1. Prove the following identities:

   (a) $\cos z = \prod_{j=1}^{\infty} \left(1 - \frac{4z^2}{(2j-1)^2 \pi^2} \right)$,

   (b) $\sinh z = z \prod_{j=1}^{\infty} \left(1 + \frac{z^2}{j^2 \pi^2} \right)$,

(c) $\cosh z = \prod_{j=1}^{\infty} \left(1 + \frac{4z^2}{(2j-1)^2\pi^2} \right)$.

2.10   Blaschke Products

Suppose that $\{z_n\}$ is a sequence S of complex numbers such that $\lim_{n\to\infty} |z_n| = \infty$. Then Weierstrass's theorem gives an entire function whose zeros are the terms of S. In this section we consider functions that are analytic on an open disc centred at the origin. We describe a condition on a sequence of complex numbers that guarantees the existence of such a function whose zeros are the terms of the sequence. We consider first the case where the disc is the unit disc and the terms of the sequence are nonzero.

Given a sequence $S = (z_1, z_2, \ldots)$ of nonzero numbers in the disc $\mathcal{D} = D(0; 1)$, a **Blaschke product** on $\mathcal{D}$ associated with S is defined as a function of the form

$$B(z) = e^{i\alpha} \prod_{j=1}^{\infty} b_j(z),$$

where

$$b_j(z) = \frac{|z_j|}{z_j} \cdot \frac{z - z_j}{\overline{z_j}z - 1}$$

for all j. Often α is taken to be 0. Note that $\overline{z_j}z - 1 \neq 0$ on $\mathcal{D}$, since $|z_j| < 1$ and $|z| < 1$. Moreover the zeros of $B(z)$ are precisely the terms of S. Observe also that the zeros of $z^m B(z)$ are the terms of S together with a zero of multiplicity m at 0. The factor $|z_j|/z_j$ is inserted for normalization.

The proof of the next theorem uses the maximum modulus theorem. The special case of the maximum modulus theorem that we need is the following result. Let f be an analytic function on $\mathcal{D} = D(0; R)$ and suppose that there exists $w \in \mathcal{D}$ such that $|f(z)| \leq |f(w)|$ for all $z \in \mathcal{D}$. Then f is a constant function. Furthermore, if f is continuous on the closed disc $\overline{\mathcal{D}} = \overline{D}(0; R)$, then $|f(z)|$ attains its maximum on the boundary $\overline{\mathcal{D}} - \mathcal{D}$.

Theorem 2.10.1 *Let $S = (z_1, z_2, \ldots)$ be a sequence of nonzero complex numbers in the unit open disc $\mathcal{D}$. Then the Blaschke product on $\mathcal{D}$ associated with S is absolutely convergent if and only if the series $\sum_{j=1}^{\infty}(1 - |z_j|)$ converges, and in that case the Blaschke product defines an analytic function B on $\mathcal{D}$ such that $|B(z)| \leq 1$ for all $z \in \mathcal{D}$.*

Proof First note that

$$\prod_{j=1}^{\infty} b_j(z) = \prod_{j=1}^{\infty} (1 + (b_j(z) - 1))$$

for all $z \in \mathcal{D}$. Therefore, by Theorem 2.2.8, the Blaschke product converges absolutely if and only if $\sum_{j=1}^{\infty} |1 - b_j(z)|$ converges.

Let $|z| \le r < 1$. Since $z_j \in \mathcal{D}$ for each j, we have $|z_j||z| < 1$. Consequently

$$
\begin{aligned}
|1 - b_j(z)| &= \left| 1 - \frac{|z_j|(z - z_j)}{z_j(\overline{z_j}z - 1)} \right| \\
&= \left| \frac{z_j(\overline{z_j}z - 1) - |z_j|(z - z_j)}{z_j(1 - \overline{z_j}z)} \right| \\
&= \left| \frac{|z_j|^2 z - z_j - |z_j|z + |z_j|z_j}{z_j(1 - \overline{z_j}z)} \right| \\
&= \left| \frac{(z_j + |z_j|z)(|z_j| - 1)}{z_j(1 - \overline{z_j}z)} \right| \\
&\le \frac{|z_j| + |z_j||z|}{|z_j||1 - |z_j||z||}(1 - |z_j|) \\
&= \frac{1 + |z|}{1 - |z_j||z|}(1 - |z_j|) \\
&\le \frac{1 + r}{1 - r}(1 - |z_j|).
\end{aligned}
$$

We conclude by the Weierstrass M-test that if $\sum_{j=1}^{\infty}(1 - |z_j|)$ converges then the series $\sum_{j=1}^{\infty} |b_j(z) - 1|$ is uniformly convergent on every compact subset of $\mathcal{D}$. Therefore, by Corollary 2.4.5, the Blaschke product converges to a function that is analytic on $\mathcal{D}$.

Conversely, suppose the Blaschke product converges absolutely on $\mathcal{D}$. Then the series $\sum_{j=1}^{\infty} |1 - b_j(z)|$ converges. As before,

$$
\begin{aligned}
|1 - b_j(z)| &= \frac{\left| 1 + \frac{|z_j|}{z_j}|z| \right|}{|1 - \overline{z_j}z|}(1 - |z_j|) \\
&\ge \frac{1 - |z|}{1 + |z_j||z|}(1 - |z_j|) \\
&\ge \frac{1 - r}{2}(1 - |z_j|).
\end{aligned}
$$

Hence $\sum_{j=1}^{\infty}(1 - |z_j|)$ converges by the comparison test.

The partial product

$$B_n(z) = \prod_{j=1}^{n} b_j(z)$$

is analytic on $\mathcal{D}$ and continuous on $\overline{D}(0; 1)$. It is easy to verify that if $|z| = 1$ then $|b_j(z)| = 1$ for all j. Hence $|B_n(z)| = 1$ for all n in this case. The maximum modulus theorem shows that $|B_n(z)| < 1$ if $|z| < 1$. Therefore $|B(z)| \le 1$ on $\mathcal{D}$. $\square$

Remark It follows from Theorem 2.2.2 that the convergence of $\sum_{j=1}^{\infty}(1 - |z_j|)$ is equivalent to that of $\prod_{j=1}^{\infty}|z_j|$. We sometimes refer to this necessary and sufficient condition for the absolute convergence of the Blaschke product as the Blaschke condition.

In regard to the disc $D(0; R)$ of radius $R > 0$ centred at the origin, we replace z_j and z by $w_j = z_j/R$ and $w = z/R$, respectively, and consider the infinite product

$$\prod_{j=1}^{\infty} \frac{|w_j|(w - w_j)}{w_j(\overline{w_j}w - 1)}.$$

We then obtain the following result from Theorem 2.10.1.

Theorem 2.10.2 *Let $S = (z_1, z_2, \ldots)$ be a sequence of nonzero numbers in $D(0; R)$. Then the product*

$$\prod_{j=1}^{\infty} \frac{R|z_j|(z - z_j)}{z_j(\overline{z_j}z - R^2)}$$

converges absolutely on $D(0; R)$ if and only if $\sum_{j=1}^{\infty}(R - |z_j|)$ converges, and in that case the product defines a function that is analytic on $D(0; R)$.

We now show that if f is a nonzero analytic bounded function defined on an open disc and its zeros satisfy the necessary condition, then f can be represented in terms of Blaschke products.

Theorem 2.10.3 *Let f be a nonzero bounded analytic function defined on the disc $D(0; R)$. Suppose that f has a zero of multiplicity m at 0 and a sequence $(z_1, z_2, \ldots)$ of nonzero zeros in which each zero appears as many times as its multiplicity. If $\sum_{j=1}^{\infty}(R - |z_j|)$ converges, then f can be written as*

$$f(z) = g(z)\left(\frac{z}{R}\right)^m \prod_{j=1}^{\infty} \frac{R|z_j|(z - z_j)}{z_j(\overline{z_j}z - R^2)}$$

for some analytic function g with no zeros in $D(0; R)$. Moreover, if $|f(z)| \le M$ for all $z \in D(0; R)$, then $|g(z)| \le M$ for all $z \in D(0; R)$.

Proof Let

$$B(z) = \left(\frac{z}{R}\right)^m \prod_{j=1}^{\infty} \frac{R|z_j|(z - z_j)}{z_j(\overline{z_j}z - R^2)}$$

for all $z \in D(0; R)$. Then the functions f and B have precisely the same zeros, with corresponding multiplicities, in $D(0; R)$. Thus the function $Q = f/B$ can be extended to an analytic function g that has no zeros in $D(0; R)$. It follows that $f(z) = g(z)B(z)$, as both sides of the equation have the same zeros. Moreover the second assertion is clear, since $|B(z)| \leq 1$. □

A study of the boundary behaviour of Blaschke products can be found in [18].

Blaschke products have been used by mathematicians for almost a century and are important in several branches of mathematics. We have only scratched the surface of this important subject. Readers are referred to [19] and [40] for further details.

Exercises 2.10

1. Show that $|b_j(z)| = 1$ whenever $|z| = 1$.
2. For which values of n does the sequence $\{1 - 1/j^n\}$ satisfy the Blaschke condition on $D(0; 1)$?
3. Show that the zeros of

$$\cos \frac{\pi}{2 - z}$$

violate the Blaschke condition on $D(0; 2)$.

2.11   Double Infinite Products

Given a double sequence $\{z_{j,k}\}$, we write

$$P_{m,n} = \prod_{j=1}^{m} \prod_{k=1}^{n} z_{j,k}.$$

If $\lim_{(m,n)\to\infty} P_{m,n}$ exists and is equal to P, then we write

$$\prod_{j,k=1}^{\infty} z_{j,k} = P.$$

We call $\prod_{j,k=1}^{\infty} z_{j,k}$ a **double infinite product** and $P_{m,n}$ a **partial product**. We say that the infinite product is **convergent** if $\lim_{(m,n)\to\infty} P_{m,n}$ is finite and nonzero;

otherwise the product is **divergent**. If $\lim_{(m,n)\to\infty} P_{m,n} = 0$, then we say that the product **diverges** to 0. In this section we will assume that $z_{j,k} \neq 0$.

In view of Theorem 2.2.4, we ask whether the convergence of the double series $\sum_{j,k=1}^{\infty} \log z_{j,k}$ is sufficient for that of the double infinite product $\prod_{j,k=1}^{\infty} z_{j,k}$. We now show that the answer is affirmative. The proof is similar to that of Theorem 2.2.4.

Theorem 2.11.1 *If $\sum_{j,k=1}^{\infty} \log z_{j,k}$ converges to S, then $\prod_{j,k=1}^{\infty} z_{j,k}$ converges to e^S.*

Proof Letting

$$S_{m,n} = \sum_{j=1}^{m} \sum_{k=1}^{n} \log z_{j,k} = \log P_{m,n} + 2k_{m,n}\pi i$$

for some integer $k_{m,n}$, we find that

$$e^{S_{m,n}} = P_{m,n}.$$

If we take the limit as $(m, n) \to \infty$, we conclude that

$$\prod_{j,k=1}^{\infty} z_{j,k} = e^S.$$

$\square$

We now investigate the connection between the double infinite product $\prod_{j,k=1}^{\infty} z_{j,k}$ defined above and the iterated infinite products $\prod_{j=1}^{\infty} \prod_{k=1}^{\infty} z_{j,k}$ and $\prod_{k=1}^{\infty} \prod_{j=1}^{\infty} z_{j,k}$. We begin with an example where the two iterated products are equal.

Example 2.11.1 If

$$z_{j,k} = 2^{-1/2^{j+k}}$$

for all j and k, then

$$\prod_{j=1}^{\infty} \prod_{k=1}^{\infty} z_{j,k} = \prod_{j=1}^{\infty} \prod_{k=1}^{\infty} 2^{-1/2^{j+k}}.$$

Note that

$$\sum_{k=1}^{\infty} \left(-\frac{1}{2^{j+k}}\right) = -\frac{1}{2^j} \sum_{k=1}^{\infty} \frac{1}{2^k} = -\frac{1}{2^j} \cdot 1 = -\frac{1}{2^j}.$$

Therefore

$$\prod_{j=1}^{\infty}\prod_{k=1}^{\infty} 2^{-1/2^{j+k}} = \prod_{j=1}^{\infty} 2^{-1/2^{j}} = \frac{1}{2},$$

since

$$\sum_{j=1}^{\infty}\left(-\frac{1}{2^{j}}\right) = -1.$$

A similar argument shows that

$$\prod_{k=1}^{\infty}\prod_{j=1}^{\infty} 2^{-\left(\frac{1}{2}\right)^{j+k}} = \frac{1}{2}.$$

$\triangle$

It is left as an exercise for the reader to verify that the iterated infinite.products $\prod_{j=1}^{\infty}\prod_{k=1}^{\infty} 1/2^{j+k}$ and $\prod_{k=1}^{\infty}\prod_{j=1}^{\infty} 1/2^{j+k}$ both diverge to 0.

We now prove a theorem that gives a sufficient condition for both of the iterated products and the double infinite product to converge to the same limit. The proof is based on [58]. We then conclude this section with four examples from the same paper. These examples show that one of the iterated products may be convergent and the other not, and if they both converge then it may be to different limits.

Theorem 2.11.2 *If any of*

$$\sum_{j,k=1}^{\infty} \log z_{j,k}, \quad \sum_{j=1}^{\infty}\sum_{k=1}^{\infty} \log z_{j,k}, \quad \sum_{k=1}^{\infty}\sum_{j=1}^{\infty} \log z_{j,k}$$

is absolutely convergent, then

$$\prod_{j,k=1}^{\infty} z_{j,k}, \quad \prod_{j=1}^{\infty}\prod_{k=1}^{\infty} z_{j,k}, \quad \prod_{k=1}^{\infty}\prod_{j=1}^{\infty} z_{j,k}$$

all converge to the same limit.

Proof The hypothesis and Theorem 1.6.8 imply that each of the series converges absolutely to the same limit S. It then follows from Theorem 2.11.1 that

$$\prod_{j,k=1}^{\infty} z_{j,k} = e^{S}.$$

For each j let

$$r_j = \sum_{k=1}^{\infty} \log z_{j,k}.$$

Therefore

$$S = \sum_{j=1}^{\infty} \sum_{k=1}^{\infty} \log z_{j,k} = \sum_{j=1}^{\infty} r_j.$$

Moreover Theorem 2.2.4 shows that

$$\prod_{k=1}^{\infty} z_{j,k} = e^{r_j}$$

for each j. Since

$$\sum_{j=1}^{\infty} \log e^{r_j} = \sum_{j=1}^{\infty} r_j = S,$$

it follows that

$$\prod_{j=1}^{\infty} \prod_{k=1}^{\infty} z_{j,k} = \prod_{j=1}^{\infty} e^{r_j} = e^S.$$

The proof that

$$\prod_{k=1}^{\infty} \prod_{j=1}^{\infty} z_{j,k} = e^S$$

is similar.   □

Example 2.11.2 Let $x > 0$ and $y > 0$. For each $j \in \mathbb{N}$ and $k \in \mathbb{N}$ define

$$z_{j,k} = \begin{cases} (xy)^{1/2^{j-1}-1} & \text{if } j = k > 1, \\ x^{1-1/2^j} & \text{if } k = j+1, \\ y^{1-1/2^k} & \text{if } j = k+1, \\ 1 & \text{otherwise.} \end{cases}$$

Thus $\prod_{k=1}^{\infty} z_{1,k} = x^{1/2}$. If $j > 1$ then

$$\prod_{k=1}^{\infty} z_{j,k} = y^{1-1/2^{j-1}}(xy)^{1/2^{j-1}-1}x^{1-1/2^j} = x^{1/2^j},$$

in agreement with the case where $j = 1$. Hence

$$\prod_{j=1}^{\infty} \prod_{k=1}^{\infty} z_{j,k} = \prod_{j=1}^{\infty} x^{1/2^j} = x.$$

Similarly

$$\prod_{k=1}^{\infty} \prod_{j=1}^{\infty} z_{j,k} = y.$$

$\triangle$

Example 2.11.3 Let $x > 0$ and for all $j \in \mathbb{N}$ and $k \in \mathbb{N}$ define

$$z_{j,k} = \begin{cases} 2^{-j+1}x^{1/2^{j-1}-1} & \text{if } j = k > 1, \\ x^{1-1/2^j} & \text{if } k = j+1, \\ 2^k & \text{if } j = k+1, \\ 1 & \text{otherwise.} \end{cases}$$

As in Example 2.11.2,

$$\prod_{k=1}^{\infty} z_{j,k} = x^{1/2^j}.$$

However for $k > 1$ we have

$$\prod_{j=1}^{\infty} z_{j,k} = x^{1-1/2^{k-1}}2^{-k+1}x^{1/2^{k-1}-1}2^k = 2.$$

Consequently

$$\prod_{j=1}^{\infty} \prod_{k=1}^{\infty} z_{j,k} = x$$

but $\prod_{k=1}^{\infty} \prod_{j=1}^{\infty} z_{j,k}$ diverges. $\triangle$

Example 2.11.4 Let $x > 0$ and for all $j \in \mathbb{N}$ and $k \in \mathbb{N}$ define

$$z_{j,k} = \begin{cases} 2^{j-1}x^{1/2^{j-1}-1} & \text{if } j = k > 1, \\ x^{1-1/2^j} & \text{if } k = j + 1, \\ 2^{-k} & \text{if } j = k + 1, \\ 1 & \text{otherwise.} \end{cases}$$

We now have

$$\prod_{k=1}^{\infty} z_{j,k} = x^{1/2^j}$$

and

$$\prod_{j=1}^{\infty} z_{j,k} = \frac{1}{2},$$

so that

$$\prod_{j=1}^{\infty} \prod_{k=1}^{\infty} z_{j,k} = x$$

and

$$\prod_{k=1}^{\infty} \prod_{j=1}^{\infty} z_{j,k} = 0.$$

$\triangle$

Example 2.11.5 For each $j \in \mathbb{N}$ and $k \in \mathbb{N}$ define

$$z_{j,k} = \begin{cases} 2^j & \text{if } k = j + 1, \\ 2^{-k} & \text{if } j = k + 1, \\ 1 & \text{otherwise.} \end{cases}$$

Then

$$\prod_{k=1}^{\infty} z_{j,k} = 2$$

for all $j > 1$, and

$$\prod_{j=1}^{\infty} z_{j,k} = \frac{1}{2}$$

for all $k > 1$. Hence $\prod_{j=1}^{\infty} \prod_{k=1}^{\infty} z_{j,k}$ and $\prod_{k=1}^{\infty} \prod_{j=1}^{\infty} z_{j,k}$ both diverge. The latter iterated product diverges to 0. △

Exercises 2.11

1. Verify that $\prod_{j=1}^{\infty} \prod_{k=1}^{\infty} 1/2^{j+k}$ and $\prod_{k=1}^{\infty} \prod_{j=1}^{\infty} 1/2^{j+k}$ both diverge to 0.
2. For each $p > 1$ show that

$$\prod_{j,k=1}^{\infty} \frac{1}{p^{1/2^{j+k}}} = \frac{1}{p}.$$

[Hint: Apply Theorem 2.11.2.]

Chapter 3
The Gamma Function

The gamma function is a generalization of the factorial function. It is related to several other functions, including the trigonometric functions and the Riemann zeta function. This chapter is devoted to the gamma function, functions that stem directly from the gamma function such as the digamma function, and applications of these functions.

The gamma function Γ can be defined in a number of ways. It is defined in this chapter using the Weierstrass canonical representation. The construction of this infinite product is motivated by and extends the factorial function $n!$ to the positive real line and then the complex plane. There are other candidates for this extension. At the end of Sect. 2.1 Wielandt's theorem is proved. This theorem shows that Γ is the unique extension of the factorial function to the complex plane, provided the extension is bounded in a certain strip in the complex plane. The infinite product representation of the gamma function make certain identities such as Eq. (3.2.1), which relates the gamma function to $\sin \pi z$, and the duplication formula (3.2.3) easier to establish. Section 3.3 deals with the digamma and polygamma functions. These functions are closely linked to the derivatives of Γ and are of interest in their own right. We will use them in Sect. 3.5 to evaluate certain series.

Stirling's formula for Γ is derived in Sect. 3.4. This result is proved using Laplace's method coupled with the integral representation of Γ. Section 3.5 focuses on products and series that have terms which are rational in the index. It turns out that such products and series can be evaluated using the gamma and polygamma functions, and this idea leads to a number of interesting relations including a product representation for the exponential function (Theorem 3.5.2). The chapter ends with a brief discussion of another function closely related to Γ, called the beta function. The beta function can be regarded as the extension to the complex plane of the formula for binomial coefficients.

3.1   Representations of the Gamma Function

Euler [20] obtained two ways of extending the factorial function to non-integer arguments. One was by means of an improper integral which was later modified by Legendre as

$$\int_0^\infty \xi^{x-1} e^{-\xi} \, d\xi \, .$$

The other was by means of an infinite product. We choose to define the gamma function as this infinite product and then show that the product is equal to the improper integral.

Note first that

$$(k-1)! = \frac{(n+k)!}{\prod_{j=0}^n (j+k)}$$

for all non-negative integers n and positive integers k. Define

$$E(n,k) = \frac{(n+k)!}{n^k n!}$$

for all such n and k and observe that

$$\lim_{n\to\infty} E(n,k) = \lim_{n\to\infty} \frac{(n+1)(n+2)\cdots(n+k)}{n^k}$$

$$= \lim_{n\to\infty} \left(1 + \frac{1}{n}\right)\left(1 + \frac{2}{n}\right)\cdots\left(1 + \frac{k}{n}\right)$$

$$= 1.$$

Now

$$(k-1)! = \frac{(n+k)!}{n^k n!} \cdot \frac{n^k n!}{\prod_{j=0}^n (j+k)}$$

$$= E(n,k) P_n(k),$$

where

$$P_n(k) = \frac{n^k n!}{\prod_{j=0}^n (j+k)} \, .$$

Hence

$$\lim_{n \to \infty} P_n(k) = (k-1)!.$$

Define

$$\mathcal{G} = \mathbb{C} - \{0, -1, -2, \ldots\}$$

and for each $z \in \mathcal{G}$ let

$$P_n(z) = \frac{n^z n!}{\prod_{j=0}^{n}(j+z)}, \tag{3.1.1}$$

where $n^z = e^{z \log n}$.

Lemma 3.1.1 *The sequence* $\{P_n(z)\}$ *converges uniformly, on any compact subset* $\mathcal{K}$ *of* $\mathcal{G}$, *to an analytic function.*

Proof The function P_n can be recast as

$$
\begin{aligned}
P_n(z) &= \frac{n^z \prod_{j=1}^{n} j}{z \prod_{j=1}^{n}(j+z)} \\
&= \frac{n^z}{z \prod_{j=1}^{n}\left(1 + \frac{z}{j}\right)} \\
&= \frac{e^{z \log n}}{z e^z e^{\frac{z}{2}} \cdots e^{\frac{z}{n}} \prod_{j=1}^{n}\left(1 + \frac{z}{j}\right) e^{-\frac{z}{j}}} \\
&= \frac{1}{z e^{-z \log n} e^{z\left(1 + \frac{1}{2} + \ldots + \frac{1}{n}\right)} \prod_{j=1}^{n}\left(1 + \frac{z}{j}\right) e^{-\frac{z}{j}}} \\
&= \frac{1}{z e^{z h_n} \prod_{j=1}^{n}\left(1 + \frac{z}{j}\right) e^{-\frac{z}{j}}},
\end{aligned}
$$

where

$$h_n = \sum_{j=1}^{n} \frac{1}{j} - \log n.$$

Now,

$$\lim_{n \to \infty} h_n = \gamma,$$

where γ denotes the Euler constant; hence, $\{zh_n\}$ converges uniformly on any compact subset $\mathcal{K}$ of $\mathcal{G}$, by Theorem 1.4.6(2). Theorem 1.4.8 thus shows that the sequence $\{e^{zh_n}\}$ converges uniformly on $\mathcal{K}$ to a function that is bounded on $\mathcal{K}$ by Corollary 1.4.2, and therefore so does $\{ze^{zh_n}\}$.

Next, consider the infinite product

$$\prod_{j=1}^{\infty}\left(1+\frac{z}{j}\right)e^{-\frac{z}{j}},\tag{3.1.2}$$

where $z \in \mathcal{G}$. The factors of the product are nonzero for these values of z. The Maclaurin series for $e^{-z/j}$ gives

$$\left(1+\frac{z}{j}\right)e^{-\frac{z}{j}} = \left(1+\frac{z}{j}\right)\sum_{k=0}^{\infty}(-1)^k\frac{z^k}{k!j^k}$$

$$= \left(1+\frac{z}{j}\right)\left(1-\frac{z}{j}\right)+\left(1+\frac{z}{j}\right)\sum_{k=2}^{\infty}(-1)^k\frac{z^k}{k!j^k}$$

$$= 1+a_j(z),$$

where

$$a_j(z) = -\frac{z^2}{j^2}+\left(1+\frac{z}{j}\right)\sum_{k=2}^{\infty}(-1)^k\frac{z^k}{k!j^k}.$$

Now,

$$|a_j(z)| \leq \left|\frac{z^2}{j^2}\right|+\left|1+\frac{z}{j}\right|\left|\frac{z^2}{j^2}\right|\left|\sum_{k=0}^{\infty}(-1)^k\frac{z^k}{(k+2)!j^k}\right|$$

$$\leq \left|\frac{z^2}{j^2}\right|\left(1+\left|1+\frac{z}{j}\right|\sum_{k=0}^{\infty}\frac{|z|^k}{k!}\right)$$

$$\leq \left|\frac{z^2}{j^2}\right|\left(1+(1+|z|)e^{|z|}\right).$$

The compact set $\mathcal{K}$ is bounded. Let

$$\Lambda = \sup_{z\in\mathcal{K}}|z^2|\left(1+(1+|z|)e^{|z|}\right)$$

and

$$M_j = \frac{\Lambda}{j^2}.$$

Then, for all $j \in \mathbb{N}$ and $z \in \mathcal{K}$,

$$|a_j(z)| \le M_j.$$

Since $\sum_{j=1}^{\infty} M_j$ converges, the Weierstrass M-test implies that

$$\sum_{j=1}^{\infty} |a_j(z)|$$

is uniformly convergent on $\mathcal{K}$ to a bounded function. Theorem 2.4.4 thus implies that the infinite product (3.1.2) converges uniformly on $\mathcal{K}$ to a bounded function. It therefore follows from Theorem 1.4.6(2) that $\{1/P_n(z)\}$ converges uniformly on $\mathcal{K}$ to a function f that is continuous by Corollary 1.4.10. Moreover $f(z) \ne 0$ for all $z \in \mathcal{K}$ by the definition of convergence. Let

$$V = \inf_{z \in \mathcal{K}} |f(z)|.$$

Since $\mathcal{K}$ is compact and f is continuous, it follows that $V > 0$. The proof is now completed by appealing to Theorems 1.4.6(3) and 1.5.1.   □

The function defined by $\lim_{n \to \infty} P_n(z)$ is called the **gamma function** and denoted by Γ. The proof of Lemma 3.1.1 provides the representation

$$\Gamma(z) = \frac{1}{z e^{\gamma z} \prod_{j=1}^{\infty} \left(1 + \frac{z}{j}\right) e^{-\frac{z}{j}}} \tag{3.1.3}$$

for all $z \in \mathcal{G}$. It follows from Eq. (3.1.1) that

$$\Gamma(z+1) = z \Gamma(z) \tag{3.1.4}$$

for all $z \in \mathcal{G}$, because

$$\frac{P_n(z)}{P_n(z+1)} = \frac{n^z n!}{\prod_{j=0}^{n}(j+z)} \cdot \frac{\prod_{j=0}^{n}(j+1+z)}{n^{z+1} n!} = \frac{n+1+z}{nz} \to \frac{1}{z}$$

as $n \to \infty$, and that $\Gamma(1) = 1 = 0!$, since

$$P_1(z) = \frac{n \cdot n!}{(n+1)!} = \frac{n}{n+1} \to 1.$$

By induction it follows that

$$\Gamma(n + 1) = n! \tag{3.1.5}$$

for all $n \in \mathbb{N} \cup \{0\}$. Equation (3.1.3) implies that

$$\Gamma(z) \neq 0 \tag{3.1.6}$$

for all $z \in \mathcal{G}$.

For each $n \in \mathbb{N}$ and $z \in \mathbb{C}$, let

$$R_n(z) = \prod_{j=1}^{n} \left(1 + \frac{z}{j}\right) e^{-\frac{z}{j}}$$

and

$$R(z) = \prod_{j-1}^{\infty} \left(1 + \frac{z}{j}\right) e^{-\frac{z}{j}}.$$

Then for all $z \in \mathcal{G}$,

$$\Gamma(z) = \frac{1}{z e^{\gamma z} R(z)}.$$

The proof of Lemma 3.1.1 shows that R converges uniformly in any compact subset $\mathcal{K}$ of $\mathcal{G}$. If m is a positive integer, then it is clear that $R(-m)$ converges to 0. The product

$$R(z; -m) = \prod_{j=m+1}^{\infty} \left(1 + \frac{z}{j}\right) e^{-\frac{z}{j}}$$

converges uniformly on any compact subset $\mathcal{K}$ of $\mathcal{G} \cup \{-1, -2, \ldots, -m\}$.

Theorem 3.1.2 *For all $z \in \mathcal{G}$,*

$$\Gamma'(z) = \psi(z)\Gamma(z), \tag{3.1.7}$$

where

$$\psi(z) = -\frac{1}{z} - \gamma + \sum_{j=1}^{\infty} \frac{z}{j(j + z)}. \tag{3.1.8}$$

Moreover ψ is analytic and

$$\psi'(z) = \sum_{j=0}^{\infty} \frac{1}{(j+z)^2}.$$

Proof The function $ze^{\gamma z}$ is evidently differentiable at all z. We now establish the differentiability of R at all $z \in \mathcal{G}$, where

$$R(z) = \prod_{j=1}^{\infty} \left(1 + \frac{z}{j}\right) e^{-\frac{z}{j}} = \prod_{j=1}^{\infty} \frac{j+z}{je^{z/j}}$$

for all $z \in \mathbb{C}$.

Choose any compact subset $\mathcal{K}$ of $\mathcal{G}$ and define

$$R_n(z) = \prod_{j=1}^{n} \frac{j+z}{je^{z/j}}$$

for each $n \in \mathbb{N}$ and $z \in \mathcal{K}$. Then

$$\log R_n(z) = \sum_{j=1}^{n} \left(\log(j+z) - \log j - \frac{z}{j}\right)$$

and hence

$$R'_n(z) = R_n(z) \sum_{j=1}^{n} \left(\frac{1}{j+z} - \frac{1}{j}\right)$$

$$= -R_n(z) \sum_{j=1}^{n} \frac{z}{j(j+z)}.$$

Choose $M \in \mathbb{R}$ such that $M \geq |z|$. Since

$$\left|\frac{z}{j(j+z)}\right| = \frac{|z|}{j|j+z|}$$

and

$$|j+z| \geq j - |z| \geq j - M$$

whenever $j > M \geq |z|$, it therefore follows that

$$\left|\frac{z}{j(j+z)}\right| \leq \frac{M}{j(j-M)}.$$

Moreover the series

$$\sum_{j=M+1}^{\infty} \frac{1}{j(j-M)}$$

converges and hence, by the M-test,

$$\sum_{j=1}^{\infty} \frac{z}{j(j+z)} \qquad (3.1.9)$$

converges uniformly on $\mathcal{K}$ to a function that is bounded on $\mathcal{K}$. The sequence $\{R_n\}$ converges uniformly on $\mathcal{K}$ to a bounded function. Hence $\{R'_n\}$ converges uniformly on $\mathcal{K}$, by Theorem 1.4.6. As $\{R_n\}$ converges to R, Theorem 1.4.20 therefore shows that R is differentiable and

$$R'(z) = -R(z) \sum_{j=1}^{\infty} \frac{z}{j(j+z)}.$$

The gamma function is therefore differentiable on $\mathcal{K}$, and since

$$\log \Gamma(z) = -\log z - \gamma z - \log R(z),$$

we have

$$\frac{\Gamma'(z)}{\Gamma(z)} = -\frac{1}{z} - \gamma - \frac{R'(z)}{R(z)},$$

which gives Eq. (3.1.8). Since

$$\psi(z) = \frac{\Gamma'(z)}{\Gamma(z)}$$

for all $z \in \mathcal{K}$, it follows that ψ is analytic.

Series (3.1.9) is uniformly convergent on any compact subset $\mathcal{K}$ of $\mathcal{G}$. It follows that the series

$$\sum_{j=1}^{\infty} \frac{d}{dz} \left(\frac{z}{j(j+z)} \right) = \sum_{j=1}^{\infty} \frac{1}{(j+z)^2}$$

is also uniformly convergent on $\mathcal{K}$ by the M-test, for if

$$b = \sup\{|z| : z \in \mathcal{K}\}$$

and $j > b$ then

$$\left| \frac{1}{(j+z)^2} \right| \le \frac{1}{(j-|z|)^2} \le \frac{1}{(j-b)^2}.$$

Consequently we conclude from Theorem 1.4.14 that ψ is differentiable on $\mathcal{G}$ and

$$\psi'(z) = \frac{1}{z^2} + \sum_{j=1}^{\infty} \frac{1}{(j+z)^2} = \sum_{j=0}^{\infty} \frac{1}{(j+z)^2} \tag{3.1.10}$$

for all $z \in \mathcal{G}$. $\square$

Equation (3.1.3) is the **Weierstrass canonical representation** of Γ. The sequence defined by (3.1.1) is due to Gauss. An alternative product representation was derived by Euler. Specifically, Eq. (3.1.1) gives

$$\frac{1}{P_n(z)} = z n^{-z} \prod_{j=1}^{n} \left(1 + \frac{z}{j} \right),$$

and using the identity

$$\prod_{j=1}^{n-1} \left(1 + \frac{1}{j} \right) = \prod_{j=1}^{n-1} \frac{j+1}{j} = \frac{n}{n-1} \cdot \frac{n-1}{n-2} \cdots \frac{2}{1} = n,$$

we have

$$\frac{1}{P_n(z)} = z \prod_{j=1}^{n-1} \left(1 + \frac{1}{j} \right)^{-z} \prod_{j=1}^{n} \left(1 + \frac{z}{j} \right)$$

$$= z \left(1 + \frac{1}{n} \right)^{z} \prod_{j=1}^{n} \frac{1 + \frac{z}{j}}{\left(1 + \frac{1}{j} \right)^z};$$

hence,

$$\Gamma(z) = \frac{1}{z} \prod_{j=1}^{\infty} \frac{\left(1 + \frac{1}{j} \right)^{z}}{1 + \frac{z}{j}}. \tag{3.1.11}$$

Another (and perhaps more common) representation of the gamma function is given by an Euler integral of the second kind. In the real case, let

$$\Pi(x) = \int_0^\infty e^{-\xi} \xi^{x-1} \, d\xi$$

for all $x > 0$. First we show that the improper integral defining $\Pi(x)$ converges. Note that

$$\Pi(x) = \int_0^1 e^{-\xi} \xi^{x-1} \, d\xi + \int_1^\infty e^{-\xi} \xi^{x-1} \, d\xi. \tag{3.1.12}$$

Moreover $e^{-\xi} < 1$ whenever $\xi > 0$. Thus for all $a \in (0, 1)$ we have

$$\int_a^1 e^{-\xi} \xi^{x-1} \, d\xi \leq \int_a^1 \xi^{x-1} \, d\xi = \frac{1 - a^x}{x} \to \frac{1}{x}$$

as $a \to 0$. Therefore the first integral on the right hand side of Eq. (3.1.12) converges by the comparison test for improper integrals. In regard to the second, let

$$f(\xi) = e^{-\xi} \xi^{x+1}$$

for all $\xi > 1$. Because

$$f'(\xi) = (x + 1)\xi^x e^{-\xi} - \xi^{x+1} e^{-\xi} = \xi^x e^{-\xi}(x + 1 - \xi),$$

it follows that $f'(\xi) = 0$ if and only if $\xi = x + 1$. Since $f'(\xi) > 0$ when $\xi < x + 1$ and $f'(\xi) < 0$ when $\xi > x + 1$, the maximum value of f is reached at $x + 1 > 1$. Hence for all $n \in \mathbb{N}$ we have

$$\begin{aligned}
\int_1^n e^{-\xi} \xi^{x-1} \, d\xi &= \int_1^n f(\xi) \xi^{-2} \, d\xi \\
&\leq f(x + 1) \int_1^n \xi^{-2} \, d\xi \\
&= f(x + 1) \left(1 - \frac{1}{n} \right) \\
&\to f(x + 1)
\end{aligned}$$

as $n \to \infty$. This calculation completes the proof of the convergence of the integral defining $\Pi(x)$.

We now show that $\Gamma(x) = \Pi(x)$ for all $x > 0$ using the approach detailed in [69], p. 241. Let

$$Q_n(x) = \int_0^n \left(1 - \frac{\xi}{n} \right)^n \xi^{x-1} \, d\xi$$

for all $n \in \mathbb{N}$ and $x > 0$. The integral defining $Q_n(x)$ is improper if $x < 1$. However,

$$0 < \left(1 - \frac{\xi}{n}\right)^n \xi^{x-1} < \xi^{x-1}$$

whenever $0 < \xi < n$. Furthermore, for all $a \in (0, n)$ we have

$$\int_a^n \xi^{x-1} \, d\xi = \frac{n^x - a^x}{x} \to \frac{n^x}{x},$$

as $a \to 0$. The convergence of the integral defining $Q_n(x)$ is now confirmed by the comparison test.

The transformation $\xi = n\tau$ yields

$$Q_n(x) = n^x \int_0^1 (1 - \tau)^n \tau^{x-1} \, d\tau.$$

Integration by parts then gives

$$\int_0^1 (1 - \tau)^n \tau^{x-1} \, d\tau = \frac{(1 - \tau)^n \tau^x}{x} \bigg|_0^1 + \frac{n}{x} \int_0^1 (1 - \tau)^{n-1} \tau^x \, d\tau$$

$$= \frac{n}{x} \int_0^1 (1 - \tau)^{n-1} \tau^x \, d\tau,$$

and by induction it follows that

$$\int_0^1 (1 - \tau)^n \tau^{x-1} \, d\tau = \frac{n!}{\prod_{j=0}^{n-1}(x + j)} \int_0^1 \tau^{x+n-1} \, d\tau$$

$$= \frac{n!}{\prod_{j=0}^{n-1}(x + j)} \cdot \frac{1}{x + n};$$

hence

$$Q_n(x) = \frac{n^x n!}{\prod_{j=0}^n (x + j)}.$$

We thus have

$$\lim_{n \to \infty} Q_n(x) = \Gamma(x),$$

for all $x > 0$. We now show that

$$\lim_{n \to \infty} Q_n(x) = \int_0^\infty e^{-\xi} \xi^{x-1} \, d\xi.$$

Lemma 3.1.3 *If $\xi \geq 0$ then*

$$0 \leq e^{-\xi} - \left(1 - \frac{\xi}{n}\right)^n \leq \frac{\xi^2}{n} e^{-\xi} \tag{3.1.13}$$

for all $n > \xi$.

Proof Choose $n > \xi$, and let $q = \xi/n$. Then $0 \leq q < 1$. Now,

$$1 + q \leq e^q = \sum_{j=0}^{\infty} \frac{q^j}{j!} \leq \sum_{j=0}^{\infty} q^j = \frac{1}{1-q}.$$

Therefore

$$(1-q)^n \leq e^{-qn} \leq (1+q)^{-n}.$$

The first inequality shows that

$$e^{-\xi} - \left(1 - \frac{\xi}{n}\right)^n \geq 0,$$

and from the second we have

$$e^{-qn}(1+q)^n \leq 1.$$

Hence

$$e^{-qn} - (1-q)^n \leq e^{-qn} - e^{-qn}(1+q)^n(1-q)^n$$

$$= e^{-\xi}\left(1 - \left(1 - \frac{\xi^2}{n^2}\right)^n\right). \tag{3.1.14}$$

As $\xi^2/n^2 < 1$, Corollary 2.7.5 in [36] shows that

$$\left(1 - \frac{\xi^2}{n^2}\right)^n \geq 1 - \frac{\xi^2}{n},$$

and inequality (3.1.14) thus gives

$$e^{-\xi} - \left(1 - \frac{\xi}{n}\right)^n \leq e^{-\xi}\left(1 - \left(1 - \frac{\xi^2}{n}\right)\right) = \frac{\xi^2}{n}e^{-\xi}.$$

$\square$

Theorem 3.1.4 *If $x > 0$, then*

$$\Gamma(x) = \int_0^\infty e^{-\xi} \xi^{x-1} \, d\xi. \tag{3.1.15}$$

Proof We have

$$\int_0^\infty e^{-\xi} \xi^{x-1} \, d\xi - \Gamma(x) = \lim_{n \to \infty} \left(\int_0^n e^{-\xi} \xi^{x-1} \, d\xi - Q_n(x) \right)$$

$$= \lim_{n \to \infty} \left(\int_0^n e^{-\xi} \xi^{x-1} \, d\xi - \int_0^n \left(1 - \frac{\xi}{n} \right)^n \xi^{x-1} \, d\xi \right)$$

$$= \lim_{n \to \infty} \left(\int_0^n \left(e^{-\xi} - \left(1 - \frac{\xi}{n} \right)^n \right) \xi^{x-1} \, d\xi \right).$$

Inequality (3.1.13) implies

$$0 \le \int_0^n \left(e^{-\xi} - \left(1 - \frac{\xi}{n} \right)^n \right) \xi^{x-1} \, d\xi$$

$$\le \int_0^n \frac{\xi^{x+1}}{n} e^{-\xi} \, d\xi$$

$$\le \frac{1}{n} \int_0^\infty e^{-\xi} \xi^{x+1} \, d\xi.$$

The improper integral in the last inequality converges (to $\Pi(x+2)$). Therefore

$$\lim_{n \to \infty} \frac{1}{n} \int_0^\infty e^{-\xi} \xi^{x+1} \, d\xi = 0,$$

so that

$$\lim_{n \to \infty} \int_0^n \left(e^{-\xi} - \left(1 - \frac{\xi}{n} \right)^n \right) \xi^{x-1} \, d\xi = 0.$$

Equation (3.1.15) thus follows. □

If $x > 0$, then Γ can be defined using either the Weierstrass product or the Euler integral in lieu of the sequence defined by Eq. (3.1.1). Relation (3.1.4) can be used with the Euler integral to extend the definition of Γ to negative non-integer values of x. In fact, there is an integral representation of Γ when x is negative but not an integer. Let $m \in \mathbb{N} \cup \{0\}$ be such that $-(m+1) < x < -m$. Then it can be shown that

$$\Gamma(x) = \int_0^\infty \xi^{x-1} \left(e^{-\xi} - \sum_{j=0}^m (-1)^j \frac{\xi^j}{j!} \right) d\xi.$$

The details can be found in [69].

The gamma function is a generalization of the factorial function. It is natural to inquire whether there exist functions f, other than Γ, that satisfy $f(1) = 1$ and

$$f(z+1) = zf(z)$$

for all $z \in \mathcal{G}$. Indeed there do. For instance, the function $\Gamma(z) \cos 2\pi z$ also satisfies those conditions. Other possibilities include

$$f(z) = \frac{1}{z e^{g(z)} \prod_{j=1}^{\infty} \left(1 + \frac{z}{j}\right) e^{-\frac{z}{j}}}$$

where g is any entire function such that $g(1) = \gamma$ and

$$g(z+1) - g(z) = \gamma + 2k\pi i$$

for any integer k. In order to secure the uniqueness of the gamma function, we therefore need to hypothesize another property.

A real-valued function f whose domain includes an interval I is defined as **concave** on I if for all $x_1 \in I$ and $x_2 \in I$ we have

$$f(x_3) \leq (1-t) f(x_1) + t f(x_2)$$

whenever $x_3 = (1-t)x_1 + tx_2$ for some $t \in (0, 1)$. A sufficient condition for a function to be concave on a given interval is for its second derivative to be non-negative throughout the interval. We prepare for this result with the following lemma.

Lemma 3.1.5 *Let g be a function which is integrable and non-decreasing on an interval $[x_1, x_2]$. Let $x_3 \in (x_1, x_2)$. Then*

$$\int_{x_1}^{x_3} g \leq \frac{x_3 - x_1}{x_2 - x_1} \int_{x_1}^{x_2} g.$$

Proof Since g is non-decreasing, we have

$$(x_2 - x_3) \int_{x_1}^{x_3} g \leq (x_2 - x_3)(x_3 - x_1)g(x_3) \leq (x_3 - x_1) \int_{x_3}^{x_2} g;$$

hence

$$\frac{1}{x_2 - x_3} \int_{x_3}^{x_2} g \geq \frac{1}{x_3 - x_1} \int_{x_1}^{x_3} g,$$

so that

$$\frac{1}{x_2 - x_1} \int_{x_1}^{x_2} g = \frac{x_3 - x_1}{x_2 - x_1} \cdot \frac{1}{x_3 - x_1} \int_{x_1}^{x_3} g + \frac{x_2 - x_3}{x_2 - x_1} \cdot \frac{1}{x_2 - x_3} \int_{x_3}^{x_2} g$$

$$\geq \frac{x_3 - x_1}{x_2 - x_1} \cdot \frac{1}{x_3 - x_1} \int_{x_1}^{x_3} g + \frac{x_2 - x_3}{x_2 - x_1} \cdot \frac{1}{x_3 - x_1} \int_{x_1}^{x_3} g$$

$$= \frac{1}{x_3 - x_1} \int_{x_1}^{x_3} g$$

and the result follows immediately.   □

Theorem 3.1.6 *Let f be a function that is twice differentiable on an interval I. If $f''(x) \geq 0$ for all $x \in I$, then f is concave on I.*

Proof Let $I = [x_1, x_2]$ and choose $x_3 \in (x_1, x_2)$. The hypothesis implies that f' is non-decreasing on I. Applying Lemma 3.1.5 we therefore find that

$$f(x_3) = f(x_1) + f(x_3) - f(x_1)$$

$$= f(x_1) + \int_{x_1}^{x_3} f'$$

$$\leq f(x_1) + \frac{x_3 - x_1}{x_2 - x_1} \int_{x_1}^{x_2} f'$$

$$= f(x_1) + \frac{x_3 - x_1}{x_2 - x_1}(f(x_2) - f(x_1))$$

$$= \left(1 - \frac{x_3 - x_1}{x_2 - x_1}\right) f(x_1) + \frac{x_3 - x_1}{x_2 - x_1} f(x_2).$$

If we set

$$t = \frac{x_3 - x_1}{x_2 - x_1},$$

then it follows that $t \in (0, 1)$,

$$x_3 = (1 - t)x_1 + tx_2$$

and

$$f(x_3) \leq (1 - t)f(x_1) + tf(x_2),$$

as required.   □

A positive function f is **log-concave** on an interval I if $\log f$ is concave on I. In other words, for all $x_1 \in I$ and $x_2 \in I$ we have

$$\log f(x_3) \le (1 - t) \log f(x_1) + t \log f(x_2)$$

whenever $x_3 = (1 - t)x_1 + tx_2$ for some $t \in (0, 1)$.

We use Theorem 3.1.6 to show the log-concavity of Γ. Let $f(x) = \log \Gamma(x)$ for all $x > 0$. Then

$$f'(x) = \frac{\Gamma'(x)}{\Gamma(x)} = \psi(x),$$

so that

$$f''(x) = \psi'(x) = \sum_{j=0}^{\infty} \frac{1}{(j + x)^2} > 0.$$

We conclude that Γ is log-concave on the interval $(0, \infty)$.

We now prove the Bohr-Mollerup theorem, which asserts that Γ gives the only log-concave function $f : (0, \infty) \to \mathbb{R}$ such that $f(1) = 1$ and $f(x + 1) = xf(x)$ for all $x > 0$. Our proof is based on [33]. An alternative proof appears in [9]. In the remainder of this section we shall assume that f is a function satisfying the conditions above.

Recall that $\Gamma(x) = \lim_{n \to \infty} P_n(x)$ where

$$P_n(x) = \frac{n^x n!}{\prod_{j=0}^{n}(x + j)}.$$

Define also

$$G_n(x) = \frac{n^{x-1} n!}{\prod_{j=0}^{n-1}(x + j)}$$

and

$$H_n(x) = \frac{(n + 1)^{x-1} n!}{\prod_{j=0}^{n-1}(x + j)}.$$

Thus

$$P_n(x) = \frac{n}{x + n} G_n(x),$$

$$H_n(x) = \left(1 + \frac{1}{n}\right)^{x-1} G_n(x)$$

and P_n, G_n, H_n all converge to the gamma function.

Lemma 3.1.7 *If a > 0 then*

$$\frac{f(a+x)}{f(a)} \begin{cases} \leq a^x & \text{if } 0 \leq x \leq 1, \\ \geq a^x & \text{if } x > 1. \end{cases}$$

Proof For all $x \geq 0$ define

$$F(x) = \log f(a+x) - x \log a.$$

We shall show that F is concave.

Choose x_1 and x_2 so that $0 \leq x_1 < x_2$ and let

$$x_3 = (1-t)x_1 + tx_2$$

for some $t \in (0, 1)$. Thus

$$a + x_3 = (1-t)(a+x_1) + t(a+x_2).$$

Moreover $a + x > 0$ for all $x \geq 0$. As $\log f$ is concave, it follows that

$$\begin{aligned} F(x_3) &= \log f(a+x_3) - x_3 \log a \\ &\leq (1-t)\log f(a+x_1) + t\log f(a+x_2) - ((1-t)x_1 + tx_2)\log a \\ &= (1-t)F(x_1) + tF(x_2). \end{aligned}$$

Thus F is indeed concave.

Note that

$$\begin{aligned} F(1) &= \log f(a+1) - \log a \\ &= \log(af(a)) - \log a \\ &= \log f(a) \\ &= F(0). \end{aligned}$$

Since F is concave, it follows that $F(x) \leq \log f(a)$ for all $x \in [0, 1]$. For such x we therefore have

$$\log f(a+x) - \log f(a) \leq x \log a,$$

so that

$$\frac{f(a+x)}{f(a)} \leq a^x.$$

Suppose therefore that $x > 1$ and that $F(x) < \log f(a)$. Since $1 \in (0, x)$, concavity then forces the contradiction that $F(1) < \log f(a)$. Thus $F(x) \geq \log f(a)$ for all $x > 1$ and it follows that

$$\frac{f(a + x)}{f(a)} \geq a^x$$

in this case.   □

Corollary 3.1.8 *If $x \in [1, 2]$ then*

$$\frac{f(a + x)}{f(a)} \leq a(a + 1)^{x-1}.$$

Proof Since $x - 1 \in [0, 1]$, the lemma shows that

$$\frac{f(a + x)}{f(a + 1)} = \frac{f(a + 1 + x - 1)}{f(a + 1)} \leq (a + 1)^{x-1}.$$

Hence

$$f(a + x) \leq f(a + 1)(a + 1)^{x-1} = af(a)(a + 1)^{x-1}$$

and the result follows.   □

We are now ready to prove the Bohr-Mollerup theorem.

Theorem 3.1.9 *Let f be a log-concave real valued function, defined at all $x > 0$, such that $f(1) = 1$ and $f(x + 1) = xf(x)$ for all $x > 0$. Then $f(x) = \Gamma(x)$ for all $x > 0$.*

Proof Choose $x \in (1, 2]$. The second of the hypothesized equations implies that

$$f(x - 1) = \frac{f(x)}{x - 1} \tag{3.1.16}$$

and, by induction,

$$f(x + n) = f(x) \prod_{j=0}^{n-1} (x + j) \tag{3.1.17}$$

for any non-negative integer n. The gamma function also satisfies the hypotheses of the theorem and consequently satisfies equations (3.1.16) and (3.1.17) as well. It is therefore enough to verify the theorem for the chosen $x \in (1, 2]$.

Induction also gives

$$f(n) = (n-1)!$$

for any positive integer n. From Lemma 3.1.7 and Corollary 3.1.8 we therefore find that

$$n^x \le \frac{f(x+n)}{(n-1)!} \le n(n+1)^{x-1}.$$

Hence

$$n^{x-1}n! \le f(x+n) \le (n+1)^{x-1}n!,$$

and from Eq. (3.1.17) it follows that

$$G_n(x) = \frac{n^{x-1}n!}{\prod_{j=0}^{n-1}(x+j)} \le f(x) \le \frac{(n+1)^{x-1}n!}{\prod_{j=0}^{n-1}(x+j)} = H_n(x).$$

Since $G_n(x)$ and $H_n(x)$ both converge to $\Gamma(x)$, we infer that $f(x) = \Gamma(x)$, as required.   □

Remark Let $\mathcal{H}$ be the half plane $\{z \in \mathbb{C}: \mathrm{Re}\,(z) > 0\}$. Let f be an analytic function mapping $\mathcal{H}$ into $\mathbb{R}$ and satisfying the hypotheses of the Bohr-Mollerup theorem. We have shown that $f(x) = \Gamma(x)$ on the positive real line. Therefore $f(z) = \Gamma(z)$ on $\mathcal{H}$ by the identity theorem.

Now define

$$\Pi(z) = \int_0^\infty \xi^{z-1} e^{-\xi}\, d\xi$$

for all $z \in \mathcal{H}$. Since

$$|\Pi(z)| \le \int_0^\infty |\xi^{z-1} e^{-\xi}|\, d\xi$$

$$= \int_0^\infty \xi^{\mathrm{Re}\,(z)-1} e^{-\xi}\, d\xi, \tag{3.1.18}$$

we see that Π converges uniformly on $\mathcal{H}$. It follows that $\Pi(z) = \Gamma(z)$ for all $z \in \mathcal{H}$.

A beautiful characterization of the complex gamma function was found by Wielandt many years after the discovery of the Bohr-Mollerup theorem. It replaces the log-concavity of f with analyticity of f on $\mathcal{H}$ and boundedness on the subset of $\mathcal{H}$ where $\mathrm{Re}\,(z) \in [1, 2)$. We follow the approach of [61].

A singularity z_0 of a complex function f is **isolated** if there is an open disc $D(z_0; r)$ such that f is analytic on the punctured disc $D'(z_0; r)$. In this case f can be represented by a Laurent series on $D'(z_0; r)$. In other words, we can write

$$f(z) = \sum_{n=0}^{\infty} a_n(z - z_0)^n + \sum_{n=1}^{\infty} \frac{b_n}{(z - z_0)^n}$$

for each $z \in D'(z_0; r)$. We refer to b_1 as the **residue** of f at z_0. It may be that $b_m \neq 0$ for some positive integer m but $b_n = 0$ for all $n > m$. We then call z_0 a **pole** of **order** m. A function is **meromorphic** if its isolated singularities are poles. A pole of order 1 is said to be **simple**. In this case

$$f(z) = \sum_{n=0}^{\infty} a_n(z - z_0)^n + \frac{b_1}{z - z_0}$$

for all $z \in D'(z_0; r)$, so that

$$(z - z_0)f(z) = \sum_{n=0}^{\infty} a_n(z - z_0)^{n+1} + b_1.$$

The residue of f at z_0 is then the limit of $(z - z_0)f(z)$ as z approaches z_0.

Lemma 3.1.10 *Let f be an analytic function defined on the half plane*

$$\mathcal{H} = \{z \in \mathbb{C} \colon \operatorname{Re}(z) > 0\}$$

and satisfying the equation

$$f(z + 1) = zf(z) \tag{3.1.19}$$

for all $z \in \mathcal{H}$. Then f can be extended to a meromorphic function $\hat{f}$ on $\mathbb{C}$ that is analytic on

$$\mathcal{G} = \mathbb{C} - \{0, -1, -2, \dots\}.$$

Moreover, for each $n \in \mathbb{N}$, $-n$ is a simple pole of $\hat{f}$ and the residue of $\hat{f}$ at this pole is

$$\frac{(-1)^n f(1)}{n!}.$$

In particular, $\hat{f}$ is entire if and only if $f(1) = 0$.

Proof Choose $z \in \mathcal{G}$ and n large enough so that $\hat{z} \in \mathcal{H}$, where $\hat{z} = z + n + 1$. Define

$$\hat{f}(z) = \frac{f(\hat{z})}{\prod_{j=0}^{n}(z + j)}.$$

By induction we find from Eq. (3.1.19) that

$$f(\hat{z}) = z(z+1) \cdots (z+n) f(z).$$

Thus $\hat{f}$ is independent of n. It is an analytic function on $\mathcal{G}$ and extends f. Moreover, for each $n \in \mathbb{N}$ we have

$$\lim_{z \to -n} (z+n) \hat{f}(z) = \lim_{z \to -n} \frac{f(z+n+1)}{\prod_{j=0}^{n-1}(z+j)}$$

$$= \frac{(-1)^n}{n!} f(1).$$

The proof is complete.   □

Remark Since $\hat{f}$ is independent of n, it follows that

$$z \hat{f}(z) = z \cdot \frac{f(z+n+2)}{z \prod_{j=1}^{n+1}(z+j)}$$

$$= \frac{f(z+1+n+1)}{\prod_{j=0}^{n}(z+1+j)}$$

$$= \hat{f}(z+1).$$

Therefore $\hat{f}$ satisfies the recursive property (3.1.19).

Lemma 3.1.11 *Let s and t be entire functions such that $s(z)t(z) = 0$ for all z. Then either $s(z) = 0$ for all z or $t(z) = 0$ for all z.*

Proof Suppose that $s(z_0) \neq 0$ for some z_0. Let

$$s(z) = \sum_{j=0}^{\infty} a_j (z - z_0)^j$$

and

$$t(z) = \sum_{j=0}^{\infty} b_j (z - z_0)^j$$

for all z. Then

$$s(z)t(z) = \sum_{j=0}^{\infty} \sum_{k=0}^{j} a_{j-k} b_k (z - z_0)^j.$$

As $s(z)t(z) = 0$ for all z, it follows that

$$\sum_{k=0}^{j} a_{j-k} b_k = 0$$

for all j.

Since $s(z_0) \neq 0$ but $s(z_0)t(z_0) = 0$, we must have $t(z_0) = 0$; hence $b_0 = 0$. Assume as an inductive hypothesis that $b_j = 0$ for all $j < n$. Thus

$$0 = \sum_{k=0}^{n} a_{n-k} b_k = a_0 b_n.$$

Since $a_0 \neq 0$, we conclude that $b_n = 0$. Thus $b_j = 0$ for all j, and it follows that $t(z) = 0$ for all z.   □

Note that Γ satisfies all the hypotheses of the following theorem. (In regard to condition 2, see Eq. (3.1.18).)

Theorem 3.1.12 (Wielandt) *Let F be a function that is analytic on $\mathcal{H}$ and satisfies the following properties:*

1. *$F(z+1) = zF(z)$ for all $z \in \mathcal{H}$;*
2. *F is bounded on $\{z \in \mathbb{C} : \operatorname{Re}(z) \in [1,2)\}$;*
3. *$F(1) = 1$.*

Then $F(z) = \Gamma(z)$ for all $z \in \mathcal{H}$.

Proof Define $f(z) = F(z) - \Gamma(z)$ for each $z \in \mathcal{H}$. Then f is analytic on $\mathcal{H}$, it satisfies equation (3.1.19) and $f(1) = 0$. Lemma 3.1.10 therefore shows that it extends to an entire function $\hat{f}$. Define $g(z) = \hat{f}(z)\hat{f}(1-z)$ for all z. Thus g is an entire function.

Let

$$S = \{z \in \mathbb{C} : \operatorname{Re}(z) \in [0,1)\}.$$

We claim that g is bounded on S. Note that $\hat{f}$ satisfies condition (2) since F and Γ do so. It follows that $\hat{f}(z)$ is bounded on S because

$$\hat{f}(z) = \frac{\hat{f}(z+1)}{z}$$

for all $z \in S - \{0\}$ and $\hat{f}$ is continuous at 0. It is also bounded on $\{z \in \mathbb{C} : \operatorname{Re}(z) = 1\}$ by condition (2). Similarly $\hat{f}(1-z)$ is bounded on $\{z \in \mathbb{C} : \operatorname{Re}(z) \in [0,1]\}$, since $\operatorname{Re}(1-z) \in [0,1]$. Thus g is indeed bounded on S.

Next,

$$g(z+1) = \hat{f}(z+1)\hat{f}(-z)$$

$$= z\hat{f}(z)\hat{f}(-z)$$
$$= -\hat{f}(z)(-z)\hat{f}(-z)$$
$$= -\hat{f}(z)\hat{f}(1-z)$$
$$= -g(z)$$

for all z. It therefore follows that g is bounded on $\mathbb{C}$ and hence constant by Liouville's theorem.

Thus $g(z) = g(1) = \hat{f}(1)\hat{f}(0) = f(1)\hat{f}(0) = 0$. It therefore follows from Lemma 3.1.11 that $\hat{f}(z) = 0$ for all $z \in \mathbb{C}$. Hence $F(z) = \Gamma(z)$ for all $z \in \mathcal{H}$. □

Exercises 3.1

1. Suppose that Γ is defined by Eq. (3.1.15). Use integration by parts to prove that Γ satisfies equation (3.1.4).
2. Show that

$$\frac{1}{2}\Gamma\left(\frac{1}{2}\right) = \int_0^\infty e^{-u^2}\, du.$$

(We will see in Sect. 3.2 that $\Gamma(1/2) = \sqrt{\pi}$.)
3. Let m and n be positive integers. Use the gamma function to show that

$$\int_0^1 u^m \log^n u\, dx = \frac{(-1)^n n!}{(m+1)^{n+1}}.$$

4. Let α be a positive real number. Show that

$$\lim_{x\to\infty} \frac{\Gamma(\alpha x)}{\Gamma(x)} = \frac{1}{\alpha}.$$

3.2 Some Identities Involving the Gamma Function

In this section we establish two interesting identities for the gamma function. The first result connects Γ with the trigonometric functions.

Theorem 3.2.1 *If z is not an integer, then*

$$\Gamma(z)\Gamma(1-z) = \frac{\pi}{\sin \pi z}. \tag{3.2.1}$$

Proof The definition of Γ gives

$$\Gamma(z)\Gamma(1-z) = \lim_{n\to\infty} P_n(z) \lim_{n\to\infty} P_n(1-z)$$

$$
\begin{aligned}
&= \lim_{n \to \infty} \frac{n^z n! \cdot n^{1-z} n!}{\prod_{j=0}^{n}(z+j) \prod_{j=1}^{n+1}(j-z)} \\
&= \lim_{n \to \infty} \frac{n \prod_{j=1}^{n} j^2}{z(n+1-z) \prod_{j=1}^{n}(j+z)(j-z)} \\
&= \lim_{n \to \infty} \left(\frac{n}{n+1-z} \cdot \frac{1}{z \prod_{j=1}^{n}\left(1 - \frac{z^2}{j^2}\right)} \right) \\
&= \frac{1}{z \prod_{j=1}^{\infty}\left(1 - \frac{z^2}{j^2}\right)}.
\end{aligned}
$$

Equation (2.6.1) implies

$$
\sin \pi z = \pi z \prod_{j=1}^{\infty} \left(1 - \frac{z^2}{j^2}\right),
$$

and Eq. (3.2.1) thus follows. $\square$

Note that if $z = 1/2$, Eq. (3.2.1) gives

$$
\Gamma^2 \left(\frac{1}{2}\right) = \pi.
$$

It is clear that $\Gamma(x) > 0$ for all $x > 0$ and therefore

$$
\Gamma\left(\frac{1}{2}\right) = \sqrt{\pi}. \tag{3.2.2}
$$

Theorem 3.2.2 (Duplication Formula)   *Let* $2z \in \mathcal{G}$. *Then*

$$
2^{2z-1} \Gamma(z) \Gamma\left(z + \frac{1}{2}\right) = \sqrt{\pi} \Gamma(2z). \tag{3.2.3}
$$

Proof Note first that since $2z \in \mathcal{G}$, we have $z \in \mathcal{G}$, for if $z = -n$ for some non-negative integer n then $2z = -2n$. Similarly $z + 1/2 \in \mathcal{G}$: if $z + 1/2 = -n$ then $2z = -2n - 1$.

As $\Gamma(z) \neq 0$ for all $z \in \mathcal{G}$, it therefore follows that the function

$$
\phi(z) = \frac{2^{2z-1} \Gamma(z) \Gamma(z + \frac{1}{2})}{\Gamma(2z)}
$$

is defined. We show first that ϕ is constant and then that $\phi(z) = \sqrt{\pi}$.

The definition of Γ gives

$$\Gamma(z) = \lim_{n\to\infty} \frac{n^z n!}{\prod_{j=0}^{n}(z+j)}.$$

Since $z + 1/2 \in \mathcal{G}$ and

$$\lim_{n\to\infty} \frac{n}{z+n+\frac{1}{2}} = 1,$$

we can represent $\Gamma(z + 1/2)$ by

$$\Gamma\left(z+\frac{1}{2}\right) = \lim_{n\to\infty} \frac{n^{z+\frac{1}{2}}(n-1)!}{\prod_{j=0}^{n-1}\left(z+j+\frac{1}{2}\right)};$$

moreover,

$$\Gamma(2z) = \lim_{n\to\infty} \frac{n^{2z} n!}{\prod_{j=0}^{n}(2z+j)}.$$

Since all subsequences of a convergent sequence converge to the same limit, it follows that

$$\Gamma(2z) = \lim_{n\to\infty} \frac{(2n)^{2z}(2n)!}{\prod_{j=0}^{2n}(2z+j)}.$$

The function ϕ can thus be rewritten as

$$\phi(z) = \lim_{n\to\infty} \frac{2^{2z-1} n^{2z-\frac{1}{2}}(n!)^2 \prod_{j=0}^{2n}(2z+j)}{(2n)^{2z}(2n)! \prod_{j=0}^{n}(z+j) \prod_{j=0}^{n-1}\left(z+j+\frac{1}{2}\right)}$$

$$= \lim_{n\to\infty} \frac{(n!)^2 \prod_{j=0}^{2n}(2z+j)}{2\sqrt{n}(2n)! \prod_{j=0}^{n}(z+j) \prod_{j=0}^{n-1}\left(z+j+\frac{1}{2}\right)}.$$

Now,

$$\prod_{j=0}^{n}(z+j) = \frac{1}{2^{n+1}} \prod_{j=0}^{n}(2z+2j)$$

and

$$\prod_{j=0}^{n-1}\left(z+j+\frac{1}{2}\right)=\frac{1}{2^n}\prod_{j=0}^{n-1}(2z+2j+1);$$

since

$$\prod_{j=0}^{2n}(2z+j)=\prod_{j=0}^{n}(2z+2j)\prod_{j=0}^{n-1}(2z+2j+1),$$

it therefore follows that

$$\frac{\prod_{j=0}^{2n}(2z+j)}{\prod_{j=0}^{n}(z+j)\prod_{j=0}^{n-1}\left(z+j+\frac{1}{2}\right)}=2^{2n+1}.$$

Consequently

$$\phi(z)=\lim_{n\to\infty}\frac{(n!)^2 2^{2n}}{\sqrt{n}(2n)!}.$$

The last equation shows that ϕ is constant; therefore,

$$\phi(z)=\phi\left(\frac{1}{2}\right)=\frac{\Gamma(\frac{1}{2})\Gamma(1)}{\Gamma(1)}=\Gamma\left(\frac{1}{2}\right)=\sqrt{\pi},$$

and Eq. (3.2.3) thus follows.   □

Equation (3.2.3) is called **Legendre's duplication formula**. The relationship was generalised by Gauss, *viz.* for any integer $m>1$ and any z such that

$$\left\{z,z+\frac{1}{m},\dots,z+\frac{m-1}{m}\right\}\subset\mathcal{G},$$

we have

$$\prod_{k=0}^{m-1}\Gamma\left(z+\frac{k}{m}\right)=(2\pi)^{\frac{m-1}{2}}m^{\frac{1}{2}-mz}\Gamma(mz). \tag{3.2.4}$$

A proof of this result can be found in [69] *loc. cit.*

Exercises 3.2

1. Use Eq. (3.2.1) to derive the equation

$$\Gamma\left(\frac{1}{2}+x\right)\Gamma\left(\frac{1}{2}-x\right)=\frac{\pi}{\cos\pi x}.$$

For which values of x is this equation valid?
2. Use Eq. (3.2.1) to prove that

$$\prod_{j=1}^{8} \Gamma\left(\frac{j}{3}\right) = \frac{640}{3^6}\left(\frac{\pi}{\sqrt{3}}\right)^3.$$

3. The odd and even factorial functions are respectively defined by

$$(2n-1)!! = \prod_{j=1}^{n}(2j-1)$$

and

$$(2n)!! = \prod_{j=1}^{n}(2j)$$

for all $n \in \mathbb{N}$. Use Eq. (3.2.2) to show that

$$\Gamma\left(n + \frac{1}{2}\right) = \frac{\sqrt{\pi}}{2^n}(2n-1)!!,$$

where $n \in \mathbb{N}$.
4. Use Eq. (3.1.4) to show that

$$\Gamma(x-n) = \frac{\Gamma(x)}{\prod_{j=1}^{n}(x-j)},$$

provided x is not an integer. Use this relation to prove that

$$\Gamma\left(\frac{1}{2} - n\right) = \frac{(-1)^n 2^n \sqrt{\pi}}{(2n-1)!!}.$$

Can this equation be derived from Eq. (3.2.3)?

3.3 Analytic Functions Related to Γ

In this section we introduce the polygamma functions, which are closely related to the derivatives of Γ. The function ψ in Theorem 3.1.2 is called the **psi** or **digamma** function, and has properties and applications of interest.

Theorem 3.3.1 *For all $z \in \mathcal{G}$,*

$$\psi(z+1) = \psi(z) + \frac{1}{z}.$$ (3.3.1)

Proof Equation (3.1.8) can be rewritten

$$\psi(z) = -\frac{1}{z} - \gamma + \sum_{j=1}^{\infty}\left(\frac{1}{j} - \frac{1}{j+z}\right);$$

hence,

$$\psi(z+1) = -\frac{1}{z+1} - \gamma + \sum_{j=1}^{\infty}\left(\frac{1}{j} - \frac{1}{j+z+1}\right)$$

$$= -\frac{1}{z+1} - \gamma + \sum_{j=1}^{\infty}\left(\frac{1}{j} - \frac{1}{j+z}\right) + \sum_{j=1}^{\infty}\left(\frac{1}{j+z} - \frac{1}{j+z+1}\right)$$

$$= \psi(z) + \frac{1}{z} - \frac{1}{z+1} + \sum_{j=1}^{\infty}\left(\frac{1}{j+z} - \frac{1}{j+z+1}\right)$$

$$= \psi(z) + \frac{1}{z},$$

by Theorem 1.1.2.   □

As in the case of the gamma function, the values of the digamma function at positive integers can be determined analytically.

Corollary 3.3.2 *Let $n \in \mathbb{N}$. Then*

$$\psi(n+1) = H_n - \gamma,$$ (3.3.2)

where

$$H_n = \sum_{j=1}^{n} \frac{1}{j}.$$

Proof First,

$$\psi(1) = -1 - \gamma + \sum_{j=1}^{\infty}\left(\frac{1}{j} - \frac{1}{j+1}\right) = -\gamma.$$ (3.3.3)

Equation (3.3.2) follows inductively from Eqs. (3.3.1) and (3.3.3).   □

The numbers H_n are called **harmonic numbers**.

The digamma function can also be evaluated analytically at $1/2$. We have

$$\psi\left(\frac{1}{2}\right) = -\gamma - 2 + \sum_{j=1}^{\infty}\left(\frac{1}{j} - \frac{1}{j + \frac{1}{2}}\right)$$

$$= -\gamma - 2 + 2\sum_{j=1}^{\infty}\left(\frac{1}{2j} - \frac{1}{2j+1}\right)$$

$$= -\gamma - 2 + 2\sum_{j=2}^{\infty}\frac{(-1)^j}{j}$$

$$= -\gamma - 2\sum_{j=1}^{\infty}\frac{(-1)^{j+1}}{j}$$

$$= -\gamma - 2\log 2. \tag{3.3.4}$$

Equation (3.3.1) thus gives

$$\psi\left(\frac{2n+1}{2}\right) = 2\sum_{j=1}^{n}\frac{1}{2j-1} - \gamma - 2\log 2 \tag{3.3.5}$$

for all $n \in \mathbb{N}$, by induction. Some other specific values for the digamma function are:

$$\psi\left(\frac{1}{4}\right) = -\gamma - \frac{\pi}{2} - 3\log 2, \tag{3.3.6}$$

$$\psi\left(\frac{1}{3}\right) = -\gamma - \frac{\pi}{2\sqrt{3}} - \frac{3}{2}\log 3, \tag{3.3.7}$$

$$\psi\left(\frac{2}{3}\right) = -\gamma + \frac{\pi}{2\sqrt{3}} - \frac{3}{2}\log 3, \tag{3.3.8}$$

$$\psi\left(\frac{3}{4}\right) = -\gamma + \frac{\pi}{2} - 3\log 2 \tag{3.3.9}$$

(cf. [28], p. 945).

We saw in the proof of Theorem 3.1.2 that the series (3.1.10) is uniformly convergent on any compact subset $\mathcal{K}$ of $\mathcal{G}$. Similar arguments show that series of the form

$$\sum_{j=0}^{\infty}\frac{1}{(j+z)^{m+1}}$$

are uniformly convergent on $\mathcal{K}$ for all values of $m \in \mathbb{N}$. Hence

$$\psi^{(m)}(z) = (-1)^{m+1} m! \sum_{j=0}^{\infty} \frac{1}{(j+z)^{m+1}} \tag{3.3.10}$$

for all $m \in \mathbb{N}$ and $z \in \mathcal{K}$.

The functions $\psi^{(m)}$ are called **polygamma functions**. The function ψ' is called the **trigamma function** in some literature. The values of the polygamma functions at positive integers can be expressed in terms of the Riemann zeta function ζ. Specifically, Eq. (3.1.10) gives

$$\psi'(1) = \sum_{j=0}^{\infty} \frac{1}{(j+1)^2} = \sum_{j=1}^{\infty} \frac{1}{j^2} = \zeta(2)$$

and, for any integer $n > 1$,

$$\psi'(n) = \sum_{j=0}^{\infty} \frac{1}{(j+n)^2} = \sum_{j=n}^{\infty} \frac{1}{j^2} = \zeta(2) - \sum_{j=1}^{n-1} \frac{1}{j^2}. \tag{3.3.11}$$

Similarly, Eq. (3.3.10) gives

$$\psi^{(m)}(1) = (-1)^{m+1} m! \zeta(m+1) \tag{3.3.12}$$

for all $n \in \mathbb{N}$ and, for all $n > 1$,

$$\psi^{(m)}(n) = (-1)^{m+1} m! \left(\zeta(m+1) - \sum_{j=1}^{n-1} \frac{1}{j^{m+1}} \right). \tag{3.3.13}$$

Theorem 3.3.3 *If $x \in (-1, 1)$ then*

$$\log \Gamma(x+1) = -\gamma x + \sum_{j=2}^{\infty} \frac{(-1)^j \zeta(j)}{j} x^j. \tag{3.3.14}$$

Proof The gamma function has derivatives of all orders and $\Gamma(x+1) > 0$ whenever $x > -1$; therefore, $\log \Gamma(x+1)$ has derivatives of all orders in $(-1, 1)$ and

$$\frac{d^m}{dx^m} \log \Gamma(x+1) = \psi^{(m-1)}(x+1)$$

for all $m \geq 1$. The coefficient for x^{m+1} in a Maclaurin series for $\log \Gamma(x+1)$ would therefore be

$$\frac{1}{(m+1)!}\psi^{(m)}(1) = \frac{1}{(m+1)!}(-1)^{m+1}m!\zeta(m+1)$$

$$= \frac{(-1)^{m+1}\zeta(m+1)}{m+1}$$

for all $m > 0$. Since $\log\Gamma(1) = 0$ and $\psi(1) = -\gamma$, if $\log\Gamma(x+1)$ has a Maclaurin series then it must be given by Eq. (3.3.14). As $\Gamma(x+1) > 0$ for all $x > -1$, it follows that $\log\Gamma(x+1)$ has a Maclaurin series with radius of convergence 1. Thus Eq. (3.3.14) holds for all $x \in (-1, 1)$. □

Exercises 3.3

1. Given that Γ is differentiable on $(0, 1)$, use Eq. (3.1.4) to show that Γ is differentiable on $(-1, 0)$.
2. (a) Determine $\Gamma''(z)$ and show that

$$\Gamma''(z)\Gamma(z) > 0$$

   for all $z \in \mathcal{G}$.
   (b) Show that Γ must have a positive local minimum in the interval $(0, \infty)$.
   (c) Show that the minimum in part (b) must be in the interval $(1, 2)$.
3. (a) Use Eq. (3.1.4) to show that $\Gamma(x) < 0$ if $x \in (-2m-1, -2m)$ and $\Gamma(x) > 0$ if $x \in (-2m, -2m+1)$, where $m \in \mathbb{N} \cup \{0\}$.
   (b) For all $m \in \mathbb{N} \cup \{0\}$, show that Γ must have a local extremum in the interval $(-m-1, m)$.
   (c) If $\Gamma(x)$ has a local extremum at $x_m \in (-m-1, m)$, show that

$$|\Gamma(x_m)| \to 0$$

   as $m \to \infty$.
4. Sketch the graph of the gamma function.
5. Sketch the graph of the digamma function.
6. Prove the recurrence relation

$$\psi^{(m)}(z+1) = \psi^{(m)}(z) + (-1)^m m! z^{-m-1}.$$

3.4 Stirling's Formula

In this section we discuss an important asymptotic formula for the gamma function when x is large. Stirling's formula is a fundamental relationship that details the rate of growth of $\Gamma(x+1)$ as $x \to \infty$. This formula thus provides an approximation for $n!$ when n is large. The proof of Stirling's formula, however, is not simple, and it is necessary to introduce a technique for approximating certain integrals, Laplace's method, which in itself is of interest. The proof given here follows that in [23].

A function f is said to be **asymptotic** to a function g as $x \to \infty$ if

$$\lim_{x \to \infty} \frac{f(x)}{g(x)} = 1.$$

This relationship is denoted by

$$f \sim g$$

as $x \to \infty$. Note that no assumptions are made concerning the existence of limits for f and g individually. Indeed, the cases of interest are precisely those when these limits do not exist.

Example 3.4.1 For each $x > 0$, let $f(x) = \log x + x^2 + e^x$ and $g(x) = e^x$. Then

$$\frac{f(x)}{g(x)} = \frac{\log x}{e^x} + \frac{x^2}{e^x} + 1;$$

hence,

$$\lim_{x \to \infty} \frac{f(x)}{g(x)} = 1,$$

so that $f \sim g$ as $x \to \infty$. $\triangle$

Theorem 3.4.1 (Laplace's Method) *Let $I = [0, b)$, where b may be finite or infinite, and suppose that f is a function such that*

1. *f has a continuous second order derivative on an open interval that includes I;*
2. *f is non-decreasing on I; and*
3. *$f(0) = f'(0) = 0$ and $f''(0) > 0$.*

Let

$$L(x) = \int_0^b e^{-xf(t)} \, dt$$

for all $x \geq 0$, and suppose there exists $\alpha \geq 0$ such that $L(\alpha)$ converges. Then $L(x)$ converges for all $x > \alpha$ and

$$L(x) \sim \sqrt{\frac{\pi}{2xf''(0)}} \tag{3.4.1}$$

as $x \to \infty$.

Proof Conditions 2 and 3 imply that $f(t) \geq 0$ for all $t \in I$ and therefore

$$0 < e^{-xf(t)} \leq e^{-\alpha f(t)}$$

for all $x > \alpha$ and $t \in I$. We thus have

$$0 < L(x) \leq L(\alpha).$$

We conclude that $L(x)$ converges for all $x > \alpha$.

By hypothesis $f''(0) > 0$ and f'' is continuous on I. Thus for any $\varepsilon \in (0, f''(0))$ there exists $\delta \in (0, b)$ for which

$$|f''(t) - f''(0)| < \varepsilon \tag{3.4.2}$$

whenever $t \in [0, \delta)$. For each $x \geq \alpha$ the integral $L(x)$ can be written

$$L(x) = L_1(x) + L_2(x),$$

where

$$L_1(x) = \int_0^\delta e^{-xf(t)} \, dt$$

and

$$L_2(x) = \int_\delta^b e^{-xf(t)} \, dt.$$

Note that $L_1(x) > 0$ and $L_2(x) > 0$.

We focus first on bounds for L_1. Choose $t \in (0, \delta]$. Theorem 1.5.7 shows that there is a $\xi \in (0, \delta)$ such that

$$f(t) = f(0) + tf'(0) + t^2 \frac{f''(\xi)}{2},$$

and condition 3 therefore implies that

$$f(t) = t^2 \frac{f''(\xi)}{2}.$$

It then follows from inequality (3.4.2) applied to ξ that

$$f''(0) - \varepsilon < f''(\xi) < f''(0) + \varepsilon.$$

Thus

$$t^2 \frac{f''(0) - \varepsilon}{2} < f(t) < t^2 \frac{f''(0) + \varepsilon}{2}$$

for all $t \in (0, \delta]$ and therefore

$$e^{-xt^2 \frac{f''(0)+\varepsilon}{2}} < e^{-xf(t)} < e^{-xt^2 \frac{f''(0)-\varepsilon}{2}}$$

for all $x > 0$. Hence,

$$\int_0^\delta e^{-xt^2 \frac{f''(0)+\varepsilon}{2}} \, dt < L_1(x) < \int_0^\delta e^{-xt^2 \frac{f''(0)-\varepsilon}{2}} \, dt. \qquad (3.4.3)$$

We now construct a bound for L_2. The function f is non-decreasing on I and consequently $f(t) \geq f(\delta) \geq 0$ for all $t \in [\delta, b)$. Therefore, for all $x > 2\alpha$ we have

$$\begin{aligned}
e^{-xf(t)} &= e^{-x\frac{f(t)}{2}} e^{-x\frac{f(t)}{2}} \\
&\leq e^{-2\alpha \frac{f(t)}{2}} e^{-x\frac{f(\delta)}{2}} \\
&= e^{-\alpha f(t)} e^{-x\frac{f(\delta)}{2}}.
\end{aligned}$$

Hence

$$\begin{aligned}
L_2(x) &= \int_\delta^b e^{-xf(t)} \, dt \\
&\leq e^{-x\frac{f(\delta)}{2}} \int_\delta^b e^{-\alpha f(t)} \, dt \\
&= e^{-x\frac{f(\delta)}{2}} L_2(\alpha) \\
&< e^{-x\frac{f(\delta)}{2}} L(\alpha) \qquad (3.4.4)
\end{aligned}$$

for all $x > 2\alpha$.

Inequalities (3.4.3) and (3.4.4) therefore imply

$$\begin{aligned}
\int_0^\delta e^{-xt^2 \frac{f''(0)+\varepsilon}{2}} \, dt &< L(x) \\
&= L_1(x) + L_2(x) \\
&< \int_0^\delta e^{-xt^2 \frac{f''(0)-\varepsilon}{2}} \, dt + e^{-x\frac{f(\delta)}{2}} L(\alpha) \qquad (3.4.5)
\end{aligned}$$

for all $x > 2\alpha$ and all $\varepsilon \in (0, f''(0))$.

We now seek an asymptotic relationship as $x \to \infty$ for integrals of the form

$$h(x) = \int_0^\delta e^{-x\beta t^2} \, dt, \tag{3.4.6}$$

where $x > 0$ and β is a positive constant. Let $u = \sqrt{\beta x} t$. Then the integral (3.4.6) transforms to

$$h(x) = \int_0^{\sqrt{\beta x}\delta} \frac{e^{-u^2}}{\sqrt{\beta x}} \, du;$$

consequently,

$$\sqrt{\beta x} h(x) = \int_0^{\sqrt{\beta x}\delta} e^{-u^2} \, du. \tag{3.4.7}$$

As $x \to \infty$, $\sqrt{\beta x}\delta \to \infty$, and the integral

$$\int_0^\infty e^{-u^2} \, du$$

can be evaluated using the gamma function. Specifically, from Eqs. (3.2.2) and (3.1.15) we have

$$\sqrt{\pi} = \Gamma\left(\frac{1}{2}\right) = \int_0^\infty \frac{e^{-\xi}}{\sqrt{\xi}} \, d\xi.$$

Let $u = \sqrt{\xi}$. Then

$$\sqrt{\pi} = 2 \int_0^\infty e^{-u^2} \, du.$$

Equation (3.4.7) thus yields

$$\lim_{x \to \infty} \sqrt{\beta x} h(x) = \frac{\sqrt{\pi}}{2}, \tag{3.4.8}$$

so that

$$h(x) \sim \frac{1}{2}\sqrt{\frac{\pi}{\beta x}}.$$

Equation (3.4.8) with $\beta = (f''(0) + \varepsilon)/2$ also implies

$$\lim_{x \to \infty} \sqrt{x} \int_0^\delta e^{-xt^2 \frac{f''(0)+\varepsilon}{2}} \, dt = \lim_{x \to \infty} \sqrt{x} h(x)$$

$$= \frac{\sqrt{\pi}}{2\sqrt{\beta}}$$

$$= \sqrt{\frac{\pi}{2(f''(0) + \varepsilon)}},$$

and similarly,

$$\lim_{x \to \infty} \sqrt{x} \int_0^\delta e^{-xt^2 \frac{f''(0) - \varepsilon}{2}} \, dt = \sqrt{\frac{\pi}{2(f''(0) - \varepsilon)}},$$

since $\varepsilon < f''(0)$. In addition, note that

$$\lim_{x \to \infty} \sqrt{x} e^{-x \frac{f(\delta)}{2}} L(\alpha) = 0,$$

since

$$f(\delta) \geq f(t) > t^2 \frac{f''(0) - \varepsilon}{2} > 0$$

for all $t \in (0, \delta]$. These limits show that for each $\varepsilon \in (0, f''(0))$ there is an $X(\varepsilon)$ such that

$$\sqrt{\frac{\pi}{2(f''(0) + \varepsilon)}} - \varepsilon < \sqrt{x} \int_0^\delta e^{-xt^2 \frac{f''(0) + \varepsilon}{2}} \, dt$$

and

$$\sqrt{x} \left(\int_0^\delta e^{-xt^2 \frac{f''(0) - \varepsilon}{2}} \, dt + e^{-\frac{xf(\delta)}{2}} L(\alpha) \right) < \sqrt{\frac{\pi}{2(f''(0) - \varepsilon)}} + \varepsilon$$

for all $x > X(\varepsilon)$. For each $x > \max\{X(\varepsilon), 2\alpha\}$ it follows from inequality (3.4.5) that

$$\sqrt{\frac{\pi}{2(f''(0) + \varepsilon)}} - \varepsilon < \sqrt{x} L(x) < \sqrt{\frac{\pi}{2(f''(0) - \varepsilon)}} + \varepsilon. \tag{3.4.9}$$

Letting

$$g_1(\varepsilon_1) = \sqrt{\frac{\pi}{2(f''(0) + \varepsilon_1)}} - \varepsilon_1$$

and

$$g_2(\varepsilon_1) = \sqrt{\frac{\pi}{2(f''(0) - \varepsilon_1)}} + \varepsilon_1$$

for all $\varepsilon_1 \in (0, f''(0))$, we have

$$\lim_{\varepsilon_1 \to 0^+} g_1(\varepsilon_1) = \lim_{\varepsilon_1 \to 0^+} g_2(\varepsilon_1) = M$$

where

$$M = \sqrt{\frac{\pi}{2f''(0)}}.$$

Therefore there exists $\delta_1 > 0$ such that $|g_1(\varepsilon_1) - M| < \varepsilon$ and $|g_2(\varepsilon_1) - M| < \varepsilon$ for all ε_1 satisfying

$$0 < \varepsilon_1 < \min\{\delta_1, f''(0)\},$$

and for any such ε_1 it follows that

$$-\varepsilon < g_1(\varepsilon_1) - M < \sqrt{x}L(x) - M < g_2(\varepsilon_1) - M < \varepsilon$$

whenever $x > \max\{X(\varepsilon_1), 2\alpha\}$. We thus have

$$\lim_{x \to \infty} \sqrt{x}L(x) = \sqrt{\frac{\pi}{2f''(0)}},$$

and the result follows.   □

Example 3.4.2 Show that

$$\int_0^1 (2 - t^2)^x \, dt \sim 2^x \sqrt{\frac{\pi}{2x}}$$

as $x \to \infty$.

Solution Note that $(2 - t^2)^x = e^{-xg(t)}$, where

$$g(t) = -\log(2 - t^2)$$

for all $t \in I = [0, 1]$. The function g does not satisfy the condition $g(0) = 0$; however, we can use the function

$$f(t) = \log 2 + g(t),$$

which does satisfy $f(0) = 0$. We have

$$e^{-xf(t)} = 2^{-x}(2 - t^2)^x,$$

and therefore

$$\int_0^1 e^{-xf(t)}\, dt = 2^{-x} \int_0^1 (2 - t^2)^x\, dt.$$

Now,

$$f'(t) = \frac{2t}{2 - t^2} \geq 0$$

and

$$f''(t) = \frac{2(2 + t^2)}{(2 - t^2)^2},$$

and it is clear that f satisfies conditions 1–3 in Theorem 3.4.1 for all $t \in I$; moreover,

$$L(x) = \int_0^1 e^{-xf(t)}\, dt$$

converges for $x = 0$. Theorem 3.4.1 thus shows that

$$L(x) \sim \sqrt{\frac{\pi}{2x}}$$

for all $x > 0$, so that

$$\int_0^1 (2 - t^2)^x\, dt = 2^x L(x) \sim 2^x \sqrt{\frac{\pi}{2x}}.$$

$$\triangle$$

Theorem 3.4.2 (Stirling's Formula)

$$\Gamma(x + 1) \sim \sqrt{2\pi}\, x^{x + \frac{1}{2}} e^{-x} \tag{3.4.10}$$

as $x \to \infty$.

Proof Equation (3.1.15) implies that

$$\Gamma(x + 1) = \int_0^\infty e^{-\xi} \xi^x\, d\xi$$

for all $x > -1$. Let $\xi = x(1+t)$. If $x \neq 0$ then

$$\Gamma(x+1) = \int_{-1}^{\infty} e^{-x(1+t)}(x(1+t))^x x\, dt$$

$$= e^{-x}x^{x+1} \int_{-1}^{\infty} e^{-xt}(1+t)^x\, dt,$$

and therefore

$$e^x x^{-(x+1)}\Gamma(x+1) = L_1(x) + L_2(x), \qquad (3.4.11)$$

where

$$L_1(x) = \int_{-1}^{0} e^{-xt}(1+t)^x\, dt$$

and

$$L_2(x) = \int_{0}^{\infty} e^{-xt}(1+t)^x\, dt.$$

We use Laplace's method to find asymptotic relations for L_1 and L_2 as $x \to \infty$. We first consider L_1. Now,

$$L_1(x) = \int_{-1}^{0} e^{-xt}(1+t)^x\, dt$$

$$= \int_{0}^{1} e^{xt}(1-t)^x\, dt$$

$$= \int_{0}^{1} e^{-xf_1(t)}\, dt,$$

where

$$f_1(t) = -t - \log(1-t)$$

for all $t \in [0, 1)$. For each such t,

$$f_1'(t) = -1 + \frac{1}{1-t} = \frac{t}{1-t} \geq 0$$

and

$$f_1''(t) = \frac{1}{(1-t)^2} > 0;$$

therefore, f_1 satisfies conditions 1–3 of Theorem 3.4.1 and L_1 converges for $x = 0$. The conditions of Theorem 3.4.1 are satisfied and thus

$$L_1(x) \sim \sqrt{\frac{\pi}{2x}} \tag{3.4.12}$$

as $x \to \infty$.

A similar approach can be used to get an asymptotic relation for L_2. Here,

$$L_2(x) = \int_0^\infty e^{-xt}(1+t)^x \, dt$$
$$= \int_0^\infty e^{-x f_2(t)} \, dt,$$

where

$$f_2(t) = t - \log(1 + t).$$

For each $t \in [0, \infty)$,

$$f_2'(t) = 1 - \frac{1}{1+t} = \frac{t}{1+t} \geq 0$$

and

$$f_2''(t) = \frac{1}{(1+t)^2} > 0,$$

so that f_2 satisfies conditions 1–3 of Theorem 3.4.1. Since

$$L_2(1) = \int_0^\infty e^{-t}(1+t) \, dt = \Gamma(1) + \Gamma(2),$$

$L_2(x)$ converges for all $x \geq 1$. Theorem 3.4.1 thus gives

$$L_2(x) \sim \sqrt{\frac{\pi}{2x}}. \tag{3.4.13}$$

Relations (3.4.11)–(3.4.13) imply

$$e^x x^{-(x+1)} \Gamma(x+1) \sim 2\sqrt{\frac{\pi}{2x}} = \sqrt{\frac{2\pi}{x}},$$

which is equivalent to relation (3.4.10).   □

Relation (3.4.10) is known as **Stirling's formula**. There are several different proofs of this result in the literature. Some of the proofs are based on asymptotic expansions of $\log \Gamma$ or the digamma function, and these lead to more delicate approximations. In particular, it can be shown that

$$\Gamma(x) = \sqrt{2\pi}\, e^{-x} x^{x-\frac{1}{2}} \left(1 + \frac{1}{12x} + \frac{1}{288x^2} - \frac{139}{51840x^3} - \frac{571}{2488320x^4} + \frac{g(x)}{x^5} \right),$$

where $|g(x)|$ is bounded as $x \to \infty$. The numbers in the asymptotic series are related to the Bernoulli numbers (cf. [69], p. 253).

If x is a positive integer, then an immediate consequence of Stirling's formula is an asymptotic formula for the factorial function, *viz.*,

$$n! \sim \sqrt{2\pi}\, n^{n+\frac{1}{2}} e^{-n}. \tag{3.4.14}$$

Example 3.4.3 Use the root test (Corollary 1.2.7) to determine whether the series

$$\sum_{j=1}^{\infty} \frac{j!}{j^j}$$

converges or diverges.

Solution Let

$$a_n = \frac{n!}{n^n}$$

for all $n \in \mathbb{N}$. Evidently $a_n > 0$ and

$$a_n^{1/n} = \frac{(n!)^{1/n}}{n}.$$

Relation (3.4.14) gives

$$(n!)^{1/n} \sim \left(\sqrt{2\pi}\, n^{n+\frac{1}{2}} e^{-n} \right)^{1/n}$$
$$= (\sqrt{2\pi n})^{1/n} n e^{-1}$$
$$\sim \frac{n}{e},$$

so that

$$a_n^{1/n} = \frac{(n!)^{1/n}}{n} \sim \frac{1}{e} < 1.$$

The series therefore converges by Corollary 1.2.7.

Δ

Example 3.4.4 Use the definition of the gamma function,

$$\Gamma(x) = \lim_{n\to\infty} P_n$$

where P_n is given by Eq. (3.1.1), and Stirling's formula to prove Legendre's duplication formula, Eq. (3.2.3).

Solution Let

$$q(x) = \frac{\Gamma(x)\Gamma(x + \frac{1}{2})}{\Gamma(2x)}$$

for all x such that $2x \in \mathcal{G}$. Arguing as in the proof of Theorem 3.2.2, we obtain

$$q(x) = 2^{1-2x} \lim_{n\to\infty} \frac{(n!)^2 2^{2n}}{\sqrt{n}(2n)!}.$$

Stirling's formula gives

$$\frac{(n!)^2}{(2n)!} \sim \frac{\left(\sqrt{2\pi}n^{n+\frac{1}{2}}e^{-n}\right)^2}{\sqrt{2\pi}(2n)^{2n+\frac{1}{2}}e^{-2n}} = \frac{\sqrt{\pi n}}{2^{2n}}.$$

Hence,

$$q(x) = 2^{1-2x} \lim_{n\to\infty} \left(\frac{2^{2n}}{\sqrt{n}} \cdot \frac{\sqrt{\pi n}}{2^{2n}}\right)$$

$$= 2^{1-2x}\sqrt{\pi},$$

which implies Eq. (3.2.3).

Δ

Exercises 3.4

1. Show that

$$\lim_{n\to\infty} \left(\frac{n^n}{n!}\right)^{1/n} = e.$$

2. Show that

$$\Gamma(\alpha + x) \sim \sqrt{2\pi}x^{\alpha+x-\frac{1}{2}}e^{-x},$$

and hence

$$\frac{\Gamma(\alpha + x)}{\Gamma(\beta + x)} \sim x^{\alpha - \beta}.$$

3. Let

$$P_n = \prod_{j=1}^{n} \frac{(2j)^2}{(2j - 1)(2j + 1)}.$$

(a) Show that

$$P_n = \frac{2^{4n} (n!)^4}{(2n + 1)((2n)!)^2}.$$

(a) Use Stirling's formula and the expression above for P_n to establish the Wallis product

$$\frac{2}{1} \cdot \frac{2}{3} \cdot \frac{4}{3} \cdot \frac{4}{5} \cdot \frac{6}{5} \cdot \frac{6}{7} \cdots = \frac{\pi}{2}.$$

3.5   Applications to Products and Series

The gamma function and the polygamma functions can be used to evaluate certain types of products and series whose terms are rational functions of the index. We begin with the use of the gamma function.

Let R be a rational function and consider an infinite product of the form

$$P = \prod_{j=1}^{\infty} R(j).$$

The expression $R(j)$ can be written as

$$R(j) = C \frac{(j - a_1)(j - a_2) \cdots (j - a_m)}{(j - b_1)(j - b_2) \cdots (j - b_l)},$$

where $C, a_1, a_2, \ldots, a_m, b_1, b_2, \ldots, b_l$ are constants. If the product converges then Theorem 2.1.1 implies that $R(j) \to 1$ as $j \to \infty$. We thus require that $m = l$ and $C = 1$. Hence P can be written

$$P = \prod_{j=1}^{\infty} \frac{(j - a_1)(j - a_2) \cdots (j - a_m)}{(j - b_1)(j - b_2) \cdots (j - b_m)}. \tag{3.5.1}$$

To avoid zeros and singularities we also require that none of

$$a_1, a_2, \ldots, a_m, b_1, b_2, \ldots, b_m$$

is a positive integer.

Theorem 3.5.1 *Let P be as defined by Eq. (3.5.1). The product P is absolutely convergent if and only if*

$$\sum_{j=1}^{m} a_j = \sum_{j=1}^{m} b_j. \tag{3.5.2}$$

If P is absolutely convergent then

$$P = \prod_{j=1}^{m} \frac{\Gamma(1 - b_j)}{\Gamma(1 - a_j)}. \tag{3.5.3}$$

Proof Note first that

$$R(k) = \frac{(k - a_1)(k - a_2) \cdots (k - a_m)}{(k - b_1)(k - b_2) \cdots (k - b_m)}$$

$$= \frac{(1 - \frac{a_1}{k})(1 - \frac{a_2}{k}) \cdots (1 - \frac{a_m}{k})}{(1 - \frac{b_1}{k})(1 - \frac{b_2}{k}) \cdots (1 - \frac{b_m}{k})}$$

for each k. For each $j \in \{1, 2, \ldots, m\}$ and $k > \max\{|b_1|, |b_2|, \ldots, |b_m|\}$ we have

$$\frac{1}{1 - \frac{b_j}{k}} = 1 + \frac{b_j}{k} + \frac{1}{k^2} \sum_{l=2}^{\infty} \frac{b_j^l}{k^{l-2}} = 1 + \frac{b_j}{k} + \frac{\rho_j(k)}{k^2}$$

where

$$\rho_j(k) = \sum_{l=0}^{\infty} \frac{b_j^{l+2}}{k^l} = b_j^2 + \sum_{l=1}^{\infty} \frac{b_j^{l+2}}{k^l} \to b_j^2$$

as $k \to \infty$. Hence $\{\rho_j(k)\}$ converges for each j. Therefore

$$R(k) = \prod_{j=1}^{m} \left(1 - \frac{a_j}{k}\right)\left(1 + \frac{b_j}{k} + \frac{\rho_j(k)}{k^2}\right)$$

$$= 1 - \frac{1}{k} \left(\sum_{j=1}^{m} a_j - \sum_{j=1}^{m} b_j \right) + \frac{r(k)}{k^2},$$

where the sequence $\{r(k)\}$ converges to some number L. Hence

$$\lim_{k \to \infty} \frac{r(k)}{k} = \lim_{k \to \infty} \frac{L}{k} = 0.$$

Let

$$\alpha_k = -\frac{\mu}{k} + \frac{r(k)}{k^2},$$

where

$$\mu = \sum_{j=1}^{m} a_j - \sum_{j=1}^{m} b_j.$$

It follows that $k\alpha_k \to -\mu$ as $k \to \infty$, so that $k|\alpha_k| \to |\mu|$. Hence

$$|\alpha_k| \sim |\mu| \cdot \frac{1}{k},$$

and so $\sum_{k=1}^{\infty} |\alpha_k|$ diverges by the limit comparison test, Theorem 1.2.2(1), if $\mu \neq 0$. If $\mu = 0$, then

$$k^2 |\alpha_k| = |r(k)| \to |L|$$

as $k \to \infty$, so that

$$|\alpha_k| \sim |L| \cdot \frac{1}{k^2}.$$

As $\sum_{k=1}^{\infty} 1/k^2$ converges, so does $\sum_{k=1}^{\infty} |\alpha_k|$ in this case, by Theorem 1.2.2(1) if $L \neq 0$ and Theorem 1.2.2(2) if $L = 0$. Theorem 2.2.1 shows that P is absolutely convergent if and only if $\sum_{k=1}^{\infty} |\alpha_k|$ converges. Therefore P is absolutely convergent if and only if Eq. (3.5.2) is satisfied.

Suppose that P is absolutely convergent. Then $\mu = 0$ and hence $e^{\mu/k} = 1$ for any $k > 0$. We can thus write

$$P = \prod_{k=1}^{\infty} \frac{(1 - \frac{a_1}{k})(1 - \frac{a_2}{k}) \cdots (1 - \frac{a_m}{k})}{(1 - \frac{b_1}{k})(1 - \frac{b_2}{k}) \cdots (1 - \frac{b_m}{k})} e^{\frac{\mu}{k}}$$

$$= \prod_{k=1}^{\infty} \frac{(1 - \frac{a_1}{k})e^{\frac{a_1}{k}}(1 - \frac{a_2}{k})e^{\frac{a_2}{k}} \cdots (1 - \frac{a_m}{k})e^{\frac{a_m}{k}}}{(1 - \frac{b_1}{k})e^{\frac{b_1}{k}}(1 - \frac{b_2}{k})e^{\frac{b_2}{k}} \cdots (1 - \frac{b_m}{k})e^{\frac{b_m}{k}}}.$$

We may assume that $a_j \neq 0$ and $b_j \neq 0$ for all j, for if for example $a_j = 0$, then

$$\Gamma(1 - a_j) = \Gamma(1) = 1.$$

Thus $-a_j \in \mathcal{G}$ and $-b_j \in \mathcal{G}$, and Eq. (3.1.3) then implies that

$$\prod_{k=1}^{\infty} \left(1 - \frac{a_j}{k}\right) e^{\frac{a_j}{k}} = \frac{1}{-a_j e^{-\gamma a_j} \Gamma(-a_j)}$$

and

$$\prod_{k=1}^{\infty} \left(1 - \frac{b_j}{k}\right) e^{\frac{b_j}{k}} = \frac{1}{-b_j e^{-\gamma b_j} \Gamma(-b_j)}$$

for $j = 1, 2, \ldots, m$. Since P is absolutely convergent and

$$\lim_{k \to \infty} \left(1 - \frac{a_j}{k}\right) = \lim_{k \to \infty} \left(1 - \frac{b_j}{k}\right) = 1 > 0$$

for all j, the factors of P can be rearranged without altering the limit (Corollary 2.2.9). Hence,

$$P = \lim_{m \to \infty} \prod_{j=1}^{m} \left(\frac{e^{\gamma a_j}}{-a_j \Gamma(-a_j)} \cdot \frac{-b_j \Gamma(-b_j)}{e^{\gamma b_j}}\right)$$

$$= e^{\mu \gamma} \prod_{j=1}^{\infty} \frac{-b_j \Gamma(-b_j)}{-a_j \Gamma(-a_j)}$$

$$= \prod_{j=1}^{\infty} \frac{\Gamma(1 - b_j)}{\Gamma(1 - a_j)},$$

where we have used Eq. (3.1.4) and the fact that $\mu = 0$.   □

Example 3.5.1 Show that for all x and y such that $y, y + x, y - x \in \mathcal{G}$,

$$\prod_{j=0}^{\infty} \left(1 - \frac{x^2}{(j + y)^2}\right) = \frac{\Gamma^2(y)}{\Gamma(y + x)\Gamma(y - x)}.$$

Solution We have

$$R(j) = 1 - \frac{x^2}{(j+y)^2} = \frac{(j+y)^2 - x^2}{(j+y)^2} = \frac{(j+y+x)(j+y-x)}{(j+y)^2};$$

consequently, $a_1 = -y - x$, $a_2 = -y + x$, $b_1 = b_2 = -y$, and therefore

$$\mu = a_1 + a_2 - b_1 - b_2 = 0.$$

Theorem 3.5.1 implies

$$\prod_{j=1}^{\infty} \left(1 - \frac{x^2}{(j+y)^2}\right) = \prod_{j=1}^{2} \frac{\Gamma(1 - b_j)}{\Gamma(1 - a_j)}$$

$$= \frac{\Gamma^2(1+y)}{\Gamma(1+y+x)\Gamma(1+y-x)}$$

$$= \frac{y^2 \Gamma^2(y)}{(y^2 - x^2)\Gamma(y+x)\Gamma(y-x)}$$

$$= \frac{\Gamma^2(y)}{\left(1 - \frac{x^2}{y^2}\right)\Gamma(y+x)\Gamma(y-x)}.$$

We thus have

$$\prod_{j=0}^{\infty} \left(1 - \frac{x^2}{(j+y)^2}\right) = \left(1 - \frac{x^2}{y^2}\right) \prod_{j=1}^{\infty} \left(1 - \frac{x^2}{(j+y)^2}\right)$$

$$= \frac{\Gamma^2(y)}{\Gamma(y+x)\Gamma(y-x)}.$$

$$\triangle$$

The digamma function can be used to sum series of the form

$$\sum_{j=1}^{\infty} \frac{1}{(j+\alpha_1)(j+\alpha_2)\cdots(j+\alpha_m)}, \qquad (3.5.4)$$

where $m \geq 2$, $\alpha_k \in \mathcal{G}$ for $k = 1, 2, \ldots, m$, and $\alpha_k \neq \alpha_l$ whenever $k \neq l$. We begin with an example.

Example 3.5.2 Let

$$S = \sum_{j=1}^{\infty} \frac{1}{(j+1)(2j+1)}.$$

A partial fraction decomposition of the terms gives

$$\frac{1}{(j+1)(2j+1)} = \frac{1}{j(j+1)} - \frac{1}{j(2j+1)}.$$

The decomposition is such that the series $\sum_{j=1}^{\infty} \frac{1}{j(j+1)}$ and $\sum_{j=1}^{\infty} \frac{1}{j(2j+1)}$ converge. We thus have

$$S = \sum_{j=1}^{\infty} \frac{1}{j(j+1)} - \sum_{j=1}^{\infty} \frac{1}{j(2j+1)}.$$

Now Eq. (3.1.8) shows that

$$\sum_{j=1}^{\infty} \frac{1}{j(j+1)} = \psi(1) + 1 + \gamma$$

and

$$\sum_{j=1}^{\infty} \frac{1}{j(2j+1)} = \sum_{j=1}^{\infty} \frac{\frac{1}{2}}{j(j+\frac{1}{2})}$$

$$= \psi\left(\frac{1}{2}\right) + 2 + \gamma;$$

therefore,

$$S = \psi(1) - \psi\left(\frac{1}{2}\right) - 1$$

$$= 2\log 2 - 1,$$

where we have used Eqs. (3.3.3) and (3.3.4).   △

This example suggests a general method for summing series of the form $\sum_{j=1}^{\infty} a_j$, where

$$a_j = \frac{1}{(j+\alpha)(j+\beta)}.$$

Here it is assumed that $\alpha, \beta \in \mathcal{G}$ and $\alpha \neq \beta$. These conditions ensure that a_j has a partial fraction decomposition of the form

$$a_j = \frac{A}{j+\alpha} + \frac{B}{j+\beta},$$

where A and B are numbers such that

$$A + B = 0. \tag{3.5.5}$$

It follows that

$$-\frac{\alpha A}{j(j+\alpha)} - \frac{\beta B}{j(j+\beta)} = \frac{A}{j+\alpha} - \frac{A}{j} + \frac{B}{j+\beta} - \frac{B}{j}$$

$$= a_j - \frac{A+B}{j}$$

$$= a_j. \tag{3.5.6}$$

Let

$$b_j = \frac{\alpha}{j(j+\alpha)}$$

and

$$c_j = \frac{\beta}{j(j+\beta)}.$$

Then the series $\sum_{j=1}^{\infty} b_j$ and $\sum_{j=1}^{\infty} c_j$ converge, and

$$\sum_{j=1}^{\infty} a_j = -A \sum_{j=1}^{\infty} b_j - B \sum_{j=1}^{\infty} c_j.$$

Using Eqs. (3.1.8) and (3.3.1) we have

$$\sum_{j=1}^{\infty} b_j = \sum_{j=1}^{\infty} \frac{\alpha}{j(j+\alpha)} = \psi(\alpha) + \frac{1}{\alpha} + \gamma = \psi(\alpha+1) + \gamma,$$

and similarly

$$\sum_{j=1}^{\infty} c_j = \psi(\beta+1) + \gamma;$$

consequently,

$$\sum_{j=1}^{\infty} a_j = -A\psi(\alpha + 1) - B\psi(\beta + 1) - \gamma (A + B)$$

$$= -A\psi(1 + \alpha) - B\psi(1 + \beta),$$

where A and B satisfy equation (3.5.5).

It is clear that the approach above can be extended to series of the form (3.5.4). Suppose that the partial fraction decomposition of

$$a_j = \frac{1}{(j + \alpha_1)(j + \alpha_2) \cdots (j + \alpha_m)}$$

is

$$a_j = \sum_{k=1}^{m} \frac{A_k}{j + \alpha_k} = \frac{A_1}{j + \alpha_1} + \frac{A_2}{j + \alpha_2} + \ldots + \frac{A_m}{j + \alpha_m}.$$

The coefficient of j^{m-1} in the numerator is $\sum_{k=1}^{m} A_k = 0$. Since

$$\frac{\alpha_k A_k}{j(j + \alpha_k)} = \frac{A_k}{j} - \frac{A_k}{j + \alpha_k}$$

for each k, it therefore follows that

$$a_j = \frac{1}{j} \sum_{k=1}^{m} A_k - \sum_{k=1}^{m} \frac{\alpha_k A_k}{j(j + \alpha_k)}$$

$$= -\sum_{k=1}^{m} \frac{\alpha_k A_k}{j(j + \alpha_k)}.$$

We thus get

$$\sum_{j=1}^{\infty} \frac{1}{(j + \alpha_1)(j + \alpha_2) \cdots (j + \alpha_m)} = -\sum_{j=1}^{\infty} \sum_{k=1}^{m} \frac{\alpha_k A_k}{j(j + \alpha_k)}$$

$$= -\sum_{k=1}^{m} A_k \sum_{j=1}^{\infty} \frac{\alpha_k}{j(j + \alpha_k)}$$

$$= -\sum_{k=1}^{m} A_k (\psi(1 + \alpha_k) + \gamma)$$

$$= -\sum_{k=1}^{m} A_k \psi(1+\alpha_k) - \gamma \sum_{k=1}^{m} A_k$$

$$= -\sum_{k=1}^{m} A_k \psi(1+\alpha_k). \tag{3.5.7}$$

Example 3.5.3 Let

$$a_j = \frac{1}{(j+1)(j+3)(2j+1)}$$

$$= \frac{1}{2(j+1)(j+3)(j+\frac{1}{2})}$$

$$= \frac{1}{2}\left(\frac{-1}{j+1} + \frac{\frac{1}{5}}{j+3} + \frac{\frac{4}{5}}{j+\frac{1}{2}}\right).$$

Equation (3.5.7) yields

$$\sum_{j=1}^{\infty} \frac{1}{(j+1)(j+3)(2j+1)}$$

$$= -\frac{1}{2}\left(-\psi(2) + \frac{1}{5}\psi(4) + \frac{4}{5}\psi\left(\frac{3}{2}\right)\right)$$

$$= -\frac{1}{2}\left(\gamma - 1 + \frac{1}{5}\left(1 + \frac{1}{2} + \frac{1}{3} - \gamma\right) + \frac{4}{5}(2 - 2\log 2 - \gamma)\right)$$

$$= \frac{4}{5}\log 2 - \frac{29}{60}.$$

$\triangle$

The trigamma function can be used to extend the method above to series that have repeated linear factors in the denominators of the terms. Consider the series

$$S = \sum_{j=1}^{\infty} \frac{1}{(j+\alpha)(j+\beta)^2},$$

where $\alpha, \beta \in \mathcal{G}$ and $\alpha \neq \beta$. The conditions on α and β ensure that the terms have a partial fraction decomposition of the form

$$\frac{1}{(j+\alpha)(j+\beta)^2} = \frac{A}{j+\alpha} + \frac{B}{j+\beta} + \frac{C}{(j+\beta)^2},$$

where $A + B = 0$. Using Eq. (3.5.6) we thus have

$$\frac{1}{(j+\alpha)(j+\beta)^2} = -\frac{\alpha A}{j(j+\alpha)} - \frac{\beta B}{j(j+\beta)} + \frac{C}{(j+\beta)^2};$$

hence,

$$S = -A \sum_{j=1}^{\infty} \frac{\alpha}{j(j+\alpha)} - B \sum_{j=1}^{\infty} \frac{\beta}{j(j+\beta)} + C \sum_{j=1}^{\infty} \frac{1}{(j+\beta)^2}$$

$$= -A\psi(1+\alpha) - B\psi(1+\beta) + C\psi'(1+\beta),$$

by Eq. (3.1.10), since

$$\psi'(1+\beta) = \sum_{j=0}^{\infty} \frac{1}{(j+1+\beta)^2} = \sum_{j=1}^{\infty} \frac{1}{(j+\beta)^2}.$$

In general, suppose that the series is of the form

$$\sum_{j=1}^{\infty} \frac{1}{L(j)Q(j)},$$

where

$$L_j = \prod_{k=1}^{m} (j+\alpha_k),$$

$$Q(j) = \prod_{k=1}^{n} (j+\beta_k)^2,$$

$n \geq 1$ and $\alpha_1, \alpha_2, \ldots, \alpha_m, \beta_1, \beta_2, \ldots, \beta_n$ are distinct members of $\mathcal{G}$. Then there exists a partial fraction decomposition of the form

$$\frac{1}{L(j)Q(j)} = \sum_{k=1}^{m} \frac{A_k}{j+\alpha_k} + \sum_{k=1}^{n} \frac{B_k}{j+\beta_k} + \sum_{k=1}^{n} \frac{C_k}{(j+\beta_k)^2},$$

where the coefficient of j^{m+2n-1} in the numerator is

$$\sum_{k=1}^{m} A_k + \sum_{k=1}^{n} B_k = 0.$$

We can thus express the terms as

$$\frac{1}{L(j)Q(j)} = -\sum_{k=1}^{m} \frac{\alpha_k A_k}{j(j+\alpha_k)} - \sum_{k=1}^{n} \frac{\beta_k B_k}{j(j+\beta_k)} + \sum_{k=1}^{n} \frac{C_k}{(j+\beta_k)^2},$$

and therefore

$$\sum_{j=1}^{\infty} \frac{1}{L(j)Q(j)} = -\sum_{k=1}^{m} A_k \psi(1+\alpha_k) - \sum_{k=1}^{n} B_k \psi(1+\beta_k) + \sum_{k=1}^{n} C_k \psi'(1+\beta_k).$$

Example 3.5.4 Let

$$S = \sum_{j=1}^{\infty} \frac{1}{j^2(j+1)^2}.$$

The partial fraction decomposition is

$$\frac{1}{j^2(j+1)^2} = -\frac{2}{j} + \frac{2}{j+1} + \frac{1}{j^2} + \frac{1}{(j+1)^2}.$$

We thus have

$$\begin{aligned}
S &= 2\psi(1) - 2\psi(2) + \psi'(1) + \psi'(2) \\
&= -2\gamma - 2(1-\gamma) + \zeta(2) + \zeta(2) - 1 \\
&= \frac{\pi^2}{3} - 3.
\end{aligned}$$

$\triangle$

It should be clear at this stage that the sums of series with cubic or higher degree factors in the denominator can be written in terms of the digamma and polygamma functions. For instance, the sum of the series

$$\sum_{j=1}^{\infty} \frac{1}{j(j+1)^2(j+2)^3}$$

can be written in terms of ψ, ψ' and ψ''. The details of this extension are left to the reader. The method can even be extended to the case where there are irreducible quadratic factors in the denominator. Here the complex digamma and polygamma functions are needed. Further examples of the method are given in [1], including the complex case.

We end this section with an infinite product representation of the exponential function.

Theorem 3.5.2 *For any $z \in \mathbb{C}$ that is not a positive integer multiple of* $\log 2$,

$$e^z = \prod_{j=1}^{\infty} \left(1 - \frac{(-1)^j z}{j \log 2 - z}\right) \tag{3.5.8}$$

Proof Since $e^z = 2^{z/\log 2}$, it is sufficient to show that $P(w)$ converges to 2^w for any complex number w that is not a positive integer, where

$$P(w) = \prod_{j=1}^{\infty} \left(1 - \frac{(-1)^j w}{j - w}\right). \tag{3.5.9}$$

We assume throughout this proof that w is not a positive integer. Therefore $w/2$ and $(w - 1)/2$ are also not positive integers.

For $k \geq 1$ let

$$\begin{aligned}
c_k &= \left(1 - \frac{(-1)^{2k} w}{2k - w}\right) \left(1 - \frac{(-1)^{2k+1} w}{2k + 1 - w}\right) \\
&= \frac{2k - 2w}{2k - w} \cdot \frac{2k + 1}{2k + 1 - w} \\
&= \frac{(k - w) \left(k + \frac{1}{2}\right)}{\left(k - \frac{w}{2}\right) \left(k + \frac{1}{2} - \frac{w}{2}\right)} \\
&= \frac{(k - a_1)(k - a_2)}{(k - b_1)(k - b_2)},
\end{aligned}$$

where $a_1 = w$, $a_2 = -1/2$, $b_1 = w/2$, and $b_2 = (w-1)/2$. Thus $a_1 + a_2 = b_1 + b_2$. It therefore follows from Theorem 3.5.1 that $\prod_{k=1}^{\infty} c_k$ is absolutely convergent. Setting $v = -w/2$ and using Theorem 3.5.1, Eqs. (3.1.4) and (3.2.2) and the duplication formula (3.2.3), we therefore obtain

$$\begin{aligned}
P(w) &= \left(1 + \frac{w}{1 - w}\right) \prod_{k=1}^{\infty} c_k \\
&= \frac{1}{1 - w} \prod_{k=1}^{\infty} \frac{(k - a_1)(k - a_2)}{(k - b_1)(k - b_2)} \\
&= \frac{\Gamma(1 - b_1)\Gamma(1 - b_2)}{(1 - w)\Gamma(1 - a_1)\Gamma(1 - a_2)}
\end{aligned}$$

$$= \frac{\Gamma(1+v)\Gamma\left(\frac{3}{2}+v\right)}{(1+2v)\Gamma(1+2v)\Gamma\left(\frac{3}{2}\right)}$$

$$= \frac{v\Gamma(v) \cdot \frac{1+2v}{2}\Gamma\left(v+\frac{1}{2}\right)}{(1+2v) \cdot 2v\Gamma(2v) \cdot \frac{1}{2}\Gamma\left(\frac{1}{2}\right)}$$

$$= \frac{\Gamma(v)\Gamma\left(v+\frac{1}{2}\right)}{2\sqrt{\pi}\Gamma(2v)}$$

$$= 2^w.$$

□

Exercises 3.5

1. Let a and b be numbers such that $a, b, a+b \in \mathcal{G}$. Show that

$$\prod_{j=1}^{\infty} \frac{j(j+a+b)}{(j+a)(j+b)} = \frac{\Gamma(a+1)\Gamma(b+1)}{\Gamma(a+b+1)}.$$

2. Show that

$$\frac{\left(\Gamma(\frac{1}{4})\right)^4}{16\pi^2} = \frac{3^2}{3^2-1} \cdot \frac{5^2-1}{5^2} \cdot \frac{7^2}{7^2-1} \cdot \frac{9^2-1}{9^2} \cdots.$$

3. Prove that

$$\frac{\sqrt{\pi}}{\Gamma(1+\frac{x}{2})\Gamma(\frac{1-x}{2})} = \prod_{j=1}^{\infty}\left(1 + \frac{(-1)^j x}{j}\right),$$

provided x is neither a negative even integer nor a positive odd integer.

4. Write the sum of the series

$$\sum_{j=1}^{\infty} \frac{1}{(j+1)(4j+1)}$$

in terms of the digamma function.

5. Write the sum of the series

$$\sum_{j=1}^{\infty} \frac{1}{j(2j+1)(j+2)^2}$$

in terms of digamma and polygamma functions.

6. Write the sum of the series

$$\sum_{j=2}^{\infty} \frac{1}{(j + \frac{1}{2})^4}$$

in terms of a polygamma function.

3.6 The Beta Function

We have seen that the gamma function generalizes the factorial function: for any non-negative integer n we have $\Gamma(n+1) = n!$. In this section we introduce a function that can be used to define the binomial coefficients.

The **beta function** B is defined by the equation

$$B(z, w) = \int_0^1 \xi^{z-1}(1 - \xi)^{w-1} \, d\xi$$

where $\mathrm{Re}\,(z) > 0$ and $\mathrm{Re}\,(w) > 0$. It is sometimes called an **Euler integral of the first kind**. See [23] for a proof that the integral converges. It may appear as though this function is quite unrelated to the gamma function, but there is in fact a striking connection between the two. We shall show formally that

$$B(z, w) = \frac{\Gamma(z)\Gamma(w)}{\Gamma(z + w)}. \tag{3.6.1}$$

It follows that

$$B(n - k + 1, k + 1) = \frac{(n - k)!k!}{(n + 1)!} = \frac{1}{(n + 1)\binom{n}{k}}$$

where n and k are integers such that $n \geq k \geq 0$.

Theorem 3.6.1 *Equation (3.6.1) holds whenever* $\mathrm{Re}\,(z) > 0$ *and* $\mathrm{Re}\,(w) > 0$.

Proof Making the substitution $u = \eta\xi$, where $\eta > 0$, we obtain

$$\int_0^\infty e^{-\eta\xi} \xi^{z-1} \, d\xi = \int_0^\infty e^{-u} \frac{u^{z-1}}{\eta^{z-1}} \cdot \frac{1}{\eta} \, du$$

$$= \frac{1}{\eta^z} \int_0^\infty e^{-u} u^{z-1} \, du$$

$$= \frac{1}{\eta^z} \Gamma(z);$$

hence

$$\Gamma(z) = \int_0^\infty \eta^z e^{-\eta\xi} \xi^{z-1} \, d\xi. \qquad (3.6.2)$$

This equation holds for any $\eta > 0$. Therefore

$$\Gamma(z)\Gamma(w) = \Gamma(z) \int_0^\infty e^{-\eta} \eta^{w-1} \, d\eta$$

$$= \int_0^\infty \int_0^\infty \eta^z e^{-\eta\xi} \xi^{z-1} e^{-\eta} \eta^{w-1} \, d\xi \, d\eta$$

$$= \int_0^\infty \int_0^\infty \eta^{z+w-1} e^{-\eta(1+\xi)} \xi^{z-1} \, d\xi \, d\eta.$$

It can be shown that these integrals converge and that the order of integration may be changed. A proof would lead us too far afield, but can be found in [23]. We thus get

$$\Gamma(z)\Gamma(w) = \int_0^\infty \xi^{z-1} \int_0^\infty \eta^{z+w-1} e^{-\eta(1+\xi)} \, d\eta \, d\xi.$$

Now Eq. (3.6.2) gives

$$\Gamma(z+w) = \eta^{z+w} \int_0^\infty e^{-\eta\zeta} \zeta^{z+w-1} \, d\zeta.$$

Defining $\xi = \eta - 1 > -1$, we thus obtain

$$\frac{\Gamma(z+w)}{(1+\xi)^{z+w}} = \int_0^\infty e^{-(1+\xi)\zeta} \zeta^{z+w-1} \, d\zeta$$

$$= \int_0^\infty \eta^{z+w-1} e^{-\eta(1+\xi)} \, d\eta.$$

This equation holds for any $\xi > -1$. Therefore

$$\Gamma(z)\Gamma(w) = \int_0^\infty \xi^{z-1} \frac{\Gamma(z+w)}{(1+\xi)^{z+w}} \, d\xi$$

$$= \Gamma(z+w) \int_0^\infty \frac{\xi^{z-1}}{(1+\xi)^{z+w}} \, d\xi.$$

Finally, if we make the substitution

$$u = \frac{\xi}{1 + \xi},$$

then $u \to 0$ as $\xi \to 0$ and $u \to 1$ as $\xi \to \infty$. It follows that

$$\int_0^\infty \frac{\xi^{z-1}}{(1+\xi)^{z+w}} \, d\xi = \int_0^\infty \frac{\xi^{z-1}}{(1+\xi)^{z-1}} \cdot \frac{1}{(1+\xi)^{w-1}} \cdot \frac{1}{(1+\xi)^2} \, d\xi$$

$$= \int_0^1 u^{z-1}(1-u)^{w-1} \, du$$

$$= B(z, w),$$

and Eq. (3.6.1) follows immediately.   □

We described this proof of Eq. (3.6.1) as "formal" because we omitted the justification of the change in the order of integration. However the proof does not require any properties of the gamma function except for its expression as an integral. We can recover some of those properties, such as the duplication formula, from Eq. (3.6.1), often in a more elegant manner. For instance, letting $\xi = \sin^2 \phi$ in the definition of B, we obtain

$$B(z, w) = 2 \int_0^{\pi/2} \sin^{2z-1} \phi \cos^{2w-1} \phi \, d\phi.$$

Thus

$$B\left(\frac{1}{2}, \frac{1}{2}\right) = 2 \int_0^{\pi/2} d\phi = \pi.$$

Equation (3.6.1) therefore gives

$$\Gamma^2\left(\frac{1}{2}\right) = \Gamma(1)\pi = \pi,$$

and Eq. (3.2.2) follows.

In order to prove the duplication formula, note that

$$\frac{\Gamma^2(z)}{\Gamma(2z)} = B(z, z)$$

$$= \int_0^1 \xi^{z-1}(1-\xi)^{z-1} \, d\xi$$

$$= 2 \int_0^{1/2} (\xi(1-\xi))^{z-1} \, d\xi$$

for each z such that $\mathrm{Re}\,(z) > 0$, since the function $\xi(1 - \xi)$ is symmetric about $\xi = 1/2$. The substitution

$$\eta = (1 - 2\xi)^2$$

therefore gives

$$
\begin{aligned}
\frac{\Gamma^2(z)}{\Gamma(2z)} &= 2 \int_1^0 \left(\frac{1 - \sqrt{\eta}}{2} \cdot \frac{1 + \sqrt{\eta}}{2} \right)^{z-1} \left(-\frac{1}{4\sqrt{\eta}} \right) d\eta \\
&= 2^{1-2z} \int_0^1 \eta^{-1/2}(1 - \eta)^{z-1}\,d\eta \\
&= 2^{1-2z} B\left(\frac{1}{2}, z \right) \\
&= 2^{1-2z} \frac{\Gamma\left(\frac{1}{2} \right) \Gamma(z)}{\Gamma\left(z + \frac{1}{2} \right)}.
\end{aligned}
$$

Consequently

$$
2^{2z-1} \Gamma(z) \Gamma\left(z + \frac{1}{2} \right) = \Gamma\left(\frac{1}{2} \right) \Gamma(2z)
$$
$$
= \sqrt{\pi}\,\Gamma(2z),
$$

as required.

Chaudhry et al. [17] generalized the beta function to

$$
B(z, w; b) = \int_0^1 t^{zw}(1 - t)^{w-1} e^{-\frac{b}{t(1-t)}}\,dt,
$$

where $\mathrm{Re}\,(b) \geq 0$. They proved many interesting results. In particular they provided the following relationship between this function and the beta and gamma functions:

$$
\int_0^\infty b^{s-1} B(z, w; b)\,db = \Gamma(s) B(z + s, w + s)
$$

where the real parts of s, $z + s$ and $w + s$ are all positive. Putting $s = 1$ gives

$$
\int_0^\infty B(z, w; b)\,db = B(z + 1, w + 1),
$$

where $\mathrm{Re}\,(z) > -1$ and $\mathrm{Re}\,(w) > -1$.

Exercises 3.6

1. Establish the following recursive formulae:

$$B(z + 1, w) = \frac{z}{z + w} B(z, w),$$

$$B(z, w + 1) = \frac{w}{z + w} B(z, w).$$

2. Show that

$$B(z, w) = B(z, w + 1) + B(z + 1, w),$$

$$B(z, w + 1) = \frac{w}{z} B(z + 1, w)$$

and, for all $n \in \mathbb{N}$,

$$B(p, n + 1) = \frac{n!}{p(p + 1) \cdots (p + n)}.$$

3. Show that

$$B(x, 1 - x) = \frac{x}{\sin \pi x},$$

$$B(1 + x, 1 - x) = \frac{\pi x}{\sin \pi x}$$

and

$$B(x, y) B(x + y, 1 - y) = \frac{\pi}{x \sin \pi y}.$$

4. For all $n \in \mathbb{N}$, show that

$$B(z, w) = n \int_0^1 t^{nz-1} (1 - t^n)^{w-1} \, dt,$$

where $\mathrm{Re}\,(z) > 0$ and $\mathrm{Re}\,(w) > 0$.

5. Show that

$$\int_0^{\pi/2} \sin^n \theta \cos^m \theta \, d\theta = \frac{1}{2} B \left(\frac{n + 1}{2}, \frac{m + 1}{2} \right).$$

[Hint: let $u = \cos^2 \theta$.] Hence evaluate the integral

$$\int_0^{\pi/2} \cos^3 \theta \sin^5 \theta \, d\theta.$$

6. Prove that

$$\frac{\partial}{\partial x} B(x, y) = B(x, y) \left(\frac{\Gamma'(x)}{\Gamma(x)} - \frac{\Gamma'(x + y)}{\Gamma(x + y)} \right)$$
$$= B(x, y)(\psi(x) - \psi(x + y)),$$

where ψ is the digamma function.

7. Prove that

$$\frac{\partial^2}{\partial b^2} B(a, b) = B(a, b)((\psi(b) - \psi(a + b))^2 + \psi'(a + b))$$

and

$$\frac{\partial^2}{\partial a \partial b} B(a, b) = B(a, b)((\psi(a) - \psi(a + b))(\psi(b) - \psi(a + b)) - \psi'(a + b)).$$

8. Prove that

$$B(z, w + 1; b) + B(z + 1, w; b) = B(z, w; b).$$

Chapter 4
Prime Numbers, Partitions and Products

Identities between infinite products and series can prove elusive. For example, we know that

$$\prod_{j=1}^{\infty}\left(1 - \frac{4x^2}{(2j-1)^2\pi^2}\right) = \sum_{j=0}^{\infty}(-1)^j \frac{x^{2j}}{(2j)!}$$

for all $x \in \mathbb{R}$, because both the product and the series represent $\cos x$. The relation, however, is not obvious from the form of the product or the series, and it was not deduced by examining partial sums or products. We used the properties of the sine and cosine function along with integration to get the infinite product.

Identities between products and series are often striking. For instance, we show in Sect. 4.3 that, for $|x| < 1$,

$$\prod_{j=1}^{\infty}(1 - x^j) = 1 - x - x^2 + x^5 + x^7 - x^{12} - x^{15} + \cdots,$$

where the nonzero terms in the series have exponents of the form $(3k^2 - k)/2$ or $(3k^2 + k)/2$ for some $k \in \mathbb{N} \cup \{0\}$. Despite the simple form of the factors in the product, the proof of this identity is not straightforward.

Many identities between products and series either come from number theory or have an interpretation in number theory. In this chapter we look at a few such identities. The first section is concerned with an identity that connects prime numbers to the Riemann zeta function. The later sections are concerned with identities that come from the theory of partitions. This chapter is not meant to be comprehensive; rather, it is intended to be a short introduction to a fascinating area. References are given so that the material can be pursued in greater depth.

© The Author(s), under exclusive license to Springer Nature Switzerland AG 2022
C. H. C. Little et al., *An Introduction to Infinite Products*, SUMS Readings,
https://doi.org/10.1007/978-3-030-90646-7_4

4.1   Prime Numbers and Euler's Identity

The set $\mathbb{N} - \{1\}$ can be partitioned into two sets. If $n \in \mathbb{N} - \{1\}$ has no divisors (in $\mathbb{N}$) other than n and 1, then n is called **prime**. The set of prime numbers is denoted by $\mathcal{P}$. If $n \notin \mathcal{P}$, then it is called **composite**. For any $n \in \mathbb{N} - \{1\}$ there exist prime numbers $p_1, p_2, \ldots, p_N$ and positive integers $\alpha_1, \alpha_2, \ldots, \alpha_N$ such that

$$n = p_1^{\alpha_1} p_2^{\alpha_2} \cdots p_N^{\alpha_N}.$$

Moreover this factorization of n into positive powers of primes is unique. As an example,

$$8001 = 3^2 \cdot 7 \cdot 127.$$

This section is devoted to products of the form

$$\prod_{p \in \mathcal{P}} \left(1 - \frac{1}{p^s}\right)^{-1} \tag{4.1.1}$$

for some number $s \geq 1$, where $p \in \mathcal{P}$ indicates that the product is taken over the prime numbers. These products have applications in problems concerning the distribution of prime numbers. An explicit formula for all the prime numbers is not known, so that questions concerning the convergence of (4.1.1) must be tackled in an indirect manner. It is this indirect approach that gives this product and related series an analytical flavour distinct from the products and series studied earlier.

Observe that

$$\frac{1}{1 - \frac{1}{p^s}} = \frac{p^s}{p^s - 1} = 1 + \frac{1}{p^s - 1}.$$

Thus Theorem 2.2.1 shows that the product (4.1.1) converges if and only if the series $\sum_{p \in \mathcal{P}} 1/(p^s - 1)$ converges. As

$$\frac{1}{p^s} = \frac{1}{p^s - 1} - \frac{1}{p^s(p^s - 1)}$$

and

$$\sum_{p \in \mathcal{P}} \frac{1}{p^s(p^s - 1)}$$

converges, it follows that the product converges if and only if $\sum_{p \in \mathcal{P}} 1/p^s$ converges.

We will show that the product (4.1.1) diverges for $s = 1$. Before we embark on this demonstration, however, we need to introduce a function and establish a lemma. Define the function G by

$$G(x, m) = \prod_{p \leq x} \left(1 + \frac{1}{p} + \cdots + \frac{1}{p^m} \right) = \prod_{p \leq x} \sum_{j=0}^{m} \frac{1}{p^j}, \qquad (4.1.2)$$

where $x \in \mathbb{R}$, $x \geq 2$ and $m \in \mathbb{N}$. Here, $p \leq x$ means that the product is taken over all prime numbers less than or equal to x. For example,

$$G(6, 2) = \left(1 + \frac{1}{2} + \frac{1}{2^2} \right) \left(1 + \frac{1}{3} + \frac{1}{3^2} \right) \left(1 + \frac{1}{5} + \frac{1}{5^2} \right).$$

We also use the floor function $\lfloor x \rfloor$. Thus for any $x > 0$ we have $\lfloor x \rfloor = n$, where n is a non-negative integer, if and only if $n \leq x < n + 1$. Evidently,

$$G(x, m) = G(\lfloor x \rfloor, m).$$

Lemma 4.1.1 *For all $x \geq 2$ and $m \in \mathbb{N}$ there exists a set $\mathcal{K} \subset \mathbb{N}$ such that*

$$G(x, m) = \sum_{k \in \mathcal{K}} \frac{1}{k}, \qquad (4.1.3)$$

where $k \in \mathcal{K}$ indicates that the sum is over the elements of $\mathcal{K}$. If $2^{m+1} > x$, then $\mathcal{K}$ contains the numbers $1, 2, \ldots, \lfloor x \rfloor$.

Proof Let $p_1, p_2, \ldots, p_N$ be all the prime numbers less than or equal to x. A general term in the expansion of the product (4.1.2) is of the form

$$\frac{1}{p_1^{\alpha_1} p_2^{\alpha_2} \cdots p_N^{\alpha_N}}, \qquad (4.1.4)$$

where $\alpha_j \in \{0, 1, \ldots, m\}$ for $j = 1, 2, \ldots, N$. Conversely, all terms of this form appear in the expansion. The denominator of (4.1.4) is certainly a natural number and hence Eq. (4.1.3) is valid for some $\mathcal{K} \subset \mathbb{N}$.

Suppose that $2^{m+1} > x$ and that there is a natural number $M \notin \mathcal{K}$ such that $M \leq x$. The number M cannot have a prime divisor greater than x and therefore there exist non-negative integers $\beta_1, \beta_2, \ldots, \beta_N$ such that

$$M = p_1^{\beta_1} p_2^{\beta_2} \cdots p_N^{\beta_N}.$$

Since $M \notin \mathcal{K}$, there is a j such that the integer β_j exceeds m. Then $\beta_j \geq m + 1$ and therefore

$$M \geq p_j^{\beta_j} \geq 2^{m+1} > x,$$

which contradicts the inequality $M \leq x$. We thus conclude that the set $\mathcal{K}$ contains the numbers $1, 2, \ldots, \lfloor x \rfloor$. $\qquad\square$

Theorem 4.1.2 *The product*

$$\prod_{p \in \mathcal{P}} \left(1 - \frac{1}{p}\right)^{-1} \tag{4.1.5}$$

and the series

$$\sum_{p \in \mathcal{P}} \frac{1}{p} \tag{4.1.6}$$

diverge.

Proof For each integer $x > 1$, let

$$P(x) = \prod_{p \leq x} \frac{1}{1 - \frac{1}{p}}.$$

Since $p > 1$ for any prime number p,

$$\frac{1}{1 - \frac{1}{p}} = \sum_{j=0}^{\infty} \frac{1}{p^j} > \sum_{j=0}^{m} \frac{1}{p^j}$$

for any $m \in \mathbb{N}$; consequently,

$$P(x) = \prod_{p \leq x} \frac{1}{1 - \frac{1}{p}} > \prod_{p \leq x} \sum_{j=0}^{m} \frac{1}{p^j} = G(x, m).$$

Choose m such that $2^{m+1} > x$. Then Lemma 4.1.1 implies

$$G(x, m) \geq \sum_{j=1}^{x} \frac{1}{j}.$$

Moreover

$$\sum_{j=1}^{x} \frac{1}{j} \geq \int_{1}^{x} \frac{d\xi}{\xi},$$

and therefore

$$P(x) > \log x. \tag{4.1.7}$$

Inequality (4.1.7) implies that $P(x) \to \infty$ as $x \to \infty$ and consequently the product diverges. It follows immediately that the series also diverges.   □

Inequality (4.1.7) provides a lower bound on the growth of the function P as x increases. A bound can also be derived for the growth of the partial sums of the series. Let

$$S(x) = \sum_{p \leq x} \frac{1}{p},$$

for each $x \geq 2$. We can modify the bound in the proof of Theorem 2.3.1 to get

$$-\log\left(1 - \frac{1}{p}\right) - \frac{1}{p} = \sum_{j=2}^{\infty} \frac{1}{jp^j}$$

$$\leq \frac{1}{2} \sum_{j=2}^{\infty} \frac{1}{p^j}$$

$$= \frac{1}{2}\left(\frac{1}{1 - \frac{1}{p}} - 1 - \frac{1}{p}\right)$$

$$= \frac{1}{2}\left(\frac{p}{p-1} - \frac{p+1}{p}\right)$$

$$= \frac{1}{2p(p-1)};$$

hence,

$$\log P(x) - S(x) = \sum_{p \leq x}\left(-\log\left(1 - \frac{1}{p}\right) - \frac{1}{p}\right)$$

$$\leq \frac{1}{2} \sum_{p \leq x} \frac{1}{p(p-1)}$$

$$< \frac{1}{2} \sum_{j=2}^{\infty} \frac{1}{j(j-1)}$$

$$= \frac{1}{2} \sum_{j=2}^{\infty}\left(\frac{1}{j-1} - \frac{1}{j}\right)$$

$$= \frac{1}{2}.$$

Inequality (4.1.7) therefore implies

$$S(x) > \log \log x - \frac{1}{2}. \tag{4.1.8}$$

Inequalities (4.1.7) and (4.1.8) turn out, in fact, to capture the asymptotic behaviour of the functions P and S. A result known as Mertens's theorem shows that

$$P(x) \sim e^{\gamma} \log x \tag{4.1.9}$$

and

$$S(x) \sim \log \log x \tag{4.1.10}$$

as $x \to \infty$. We will not prove Mertens's theorem, but the reader can find a proof in [30], p. 351.

The argument above follows loosely that given by Ingham [32]. He notes that it has not been assumed that there is an infinitude of prime numbers. In fact the divergence of the product (and the series) show that there are infinitely many primes. There are of course much more elementary proofs of this result (cf. [30]).

The elements of $\mathcal{P}$ can be organized according to magnitude. Let $p_1 = 2$ and for each $n \geq 1$ let p_{n+1} be the smallest prime number such that $p_{n+1} > p_n$. The element p_n is called the nth prime number. A formula for p_n is not known. Questions concerning the distribution of prime numbers tend to be formidable. Theorem 4.1.2 does give us a little information about the distribution. Recall that the series $\sum_{j=1}^{\infty} 1/j^2$ converges whereas the series $\sum_{j=1}^{\infty} 1/p_j$ diverges. Roughly speaking this means that the natural numbers that are perfect squares are more sparsely distributed than prime numbers.

Let $s > 1$. Then $\sum_{j=1}^{\infty} 1/j^s$ converges, and it follows that $\sum_{j=1}^{\infty} 1/p_j^s$ converges. Thus the product (4.1.1) converges whenever $s > 1$. There is a remarkable relationship between this product and the Riemann zeta function. We first establish a theorem about multiplicative functions. Let f be a function defined for all $n \in \mathbb{N}$. The function f is called **completely multiplicative** if f is not identically zero and

$$f(mn) = f(m)f(n) \tag{4.1.11}$$

for all $m, n \in \mathbb{N}$.

Theorem 4.1.3 *Let f be a completely multiplicative function and suppose that the series $\sum_{j=1}^{\infty} f(j)$ is absolutely convergent. Then,*

$$\sum_{j=1}^{\infty} f(j) = \prod_{j=1}^{\infty} \frac{1}{1 - f(p_j)}, \tag{4.1.12}$$

and the product is absolutely convergent.

Proof The proof follows that given by Ingham [32].

First, if $|f(p)| \geq 1$ for some number p, then

$$|f(p^m)| = |f^m(p)| \geq 1$$

for all positive integers m, since f is multiplicative. This result contradicts the fact that $f(j) \to 0$ as $j \to \infty$, because the series converges. Hence $|f(p)| < 1$ for all p.

The absolute convergence of $\sum_{j=1}^{\infty} f(j)$ implies that for any $p \in \mathcal{P}$ the series $\sum_{j=0}^{\infty} |f(p^j)|$ converges. Let $p_a, p_b \in \mathcal{P}$. Then $\sum_{j=0}^{\infty} f(p_a^j)$ and $\sum_{j=0}^{\infty} f(p_b^j)$ are absolutely convergent and so the Cauchy product

$$\sum_{j=0}^{\infty} f(p_a^j) \sum_{j=0}^{\infty} f(p_b^j) = \sum_{k=0}^{\infty} \sum_{j=0}^{k} f(p_a^j) f(p_b^{k-j})$$

is absolutely convergent and can therefore be rearranged in any manner without changing the sum. The multiplicative property of f shows that

$$f(p_a^j) f(p_b^{k-j}) = f(p_a^j p_b^{k-j}).$$

If $p_a \neq p_b$, the Cauchy product can be arranged as a simple series of the form

$$\sum_{k \in \mathcal{K}} f(k).$$

Here $k \in \mathcal{K}$ indicates that the summation is over all k of the form $p_a^{m_1} p_b^{m_2}$ where m_1 and m_2 are non-negative integers.

Consider the product

$$F(x) = \prod_{p \leq x} \sum_{j=0}^{\infty} f(p^j),$$

where x is an integer greater than 1. This product has a finite number of factors and the argument above shows that it can be expressed in the form

$$F(x) = \sum_{k \in \mathcal{J}} f(k),$$

where the summation is over the set $\mathcal{J}$ consisting of the positive integers that have no prime factor greater than x. Let

$$G = \sum_{j=1}^{\infty} f(j).$$

Then

$$F(x) - G = - \sum_{k \in \mathbb{N} - \mathcal{J}} f(k).$$

If $k \in \mathbb{N} - \mathcal{J}$, then k must have a prime factor greater than x; consequently,

$$|F(x) - G| = \left| \sum_{k \in \mathbb{N} - \mathcal{J}} f(k) \right| \leq \sum_{k \in \mathbb{N} - \mathcal{J}} |f(k)| \leq \sum_{k=x}^{\infty} |f(k)| = \sigma(x), \qquad (4.1.13)$$

where

$$\sigma(x) = \sum_{k=x}^{\infty} |f(k)|.$$

By hypothesis $\sum_{k=1}^{\infty} |f(k)|$ converges and so $\sigma(x) \to 0$ as $x \to \infty$. Inequality (4.1.13) therefore implies that $F(x) \to G$ as $x \to \infty$. We thus have

$$\sum_{j=1}^{\infty} f(j) = \lim_{x \to \infty} F(x)$$

$$= \prod_{j=1}^{\infty} \sum_{k=0}^{\infty} f(p_j^k)$$

$$= \prod_{j=1}^{\infty} \sum_{k=0}^{\infty} f^k(p_j)$$

$$= \prod_{j=1}^{\infty} \frac{1}{1 - f(p_j)},$$

since $|f(p_j)| < 1$.

Now let

$$a_j = \frac{1}{1 - f(p_j)} - 1 = \sum_{k=1}^{\infty} f^k(p_j).$$

Then, for any n,

$$\sum_{j=1}^{n} |a_j| \le \sum_{j=1}^{n} \sum_{k=1}^{\infty} |f(p_j^k)| \le \sum_{j=1}^{\infty} |f(j)|,$$

so that the series $\sum_{j=1}^{\infty} a_j$ is absolutely convergent. The product $\prod_{j=1}^{\infty} (1 + a_j)$ is therefore absolutely convergent by Theorem 2.2.1, and hence the product in Eq. (4.1.12) is absolutely convergent. □

Corollary 4.1.4 (Euler's Identity) *If $s > 1$, then*

$$\zeta(s) = \prod_{j=1}^{\infty} \frac{1}{1 - \frac{1}{p_j^s}}. \tag{4.1.14}$$

Proof Let $f(j) = 1/j^s$. Then f is multiplicative. The series $\zeta(s) = \sum_{j=1}^{\infty} 1/j^s$ converges absolutely for all $s > 1$ and therefore Eq. (4.1.14) follows from Theorem 4.1.3. □

Euler's identity is a striking relation that connects prime numbers to the Riemann zeta function. It is the starting point for investigations into the distribution of prime numbers and in particular the number of primes no greater than a given positive number. Let $\pi(x)$ denote the number of primes less than or equal to $x \ge 2$. A central question concerns the asymptotic behaviour of $\pi(x)$ as $x \to \infty$. Chebyshev showed that for all x sufficiently large

$$\lambda \frac{x}{\log x} < \pi(x) < \frac{6}{5} \lambda \frac{x}{\log x},$$

where

$$\lambda = \frac{\log 2}{2} + \frac{\log 3}{3} + \frac{\log 5}{5} - \frac{\log 30}{30} \approx 0.92129,$$

and thus established the order of $\pi(x)$ as $x \to \infty$. Although these bounds were later refined, Chebyshev's method did not yield an asymptotic formula for $\pi(x)$. A major breakthrough came when Riemann considered ζ as a function of a complex variable. Using the (then young) theory of complex functions he was able to express $\pi(x)$ in terms of a complex integral involving ζ. Through Riemann's approach Hadamard and de la Vallée Poussin independently and almost simultaneously proved in 1896 that

$$\pi(x) \sim \frac{x}{\log x}.$$

This result is known as the prime number theorem. A proof is well beyond the scope of this book. The original proofs have been simplified substantially (cf. [32]) but nonetheless rely heavily on complex analysis. Newman's proof [71] is simpler yet but still relies on some complex analysis. Proofs that do not use complex analysis were given by Erdős and Selberg in 1949. A description of Selberg's method can be found in [30]. A short account of the rich and interesting history of the prime number theorem is given by Bateman and Diamond [11].

Exercises 4.1

1. Gauss proposed the use of the logarithmic integral

$$\mathrm{Li}(x) = \int_2^x \frac{d\xi}{\log \xi}$$

as an approximation for $\pi(x)$. Show that

$$\lim_{x \to \infty} \frac{x \mathrm{Li}(x)}{\log x} = 1.$$

2. The von Mangoldt function Λ is defined by $\Lambda(j) = \log p$ if $j = p^m$ for some prime number p and some $m \in \mathbb{N}$; otherwise $\Lambda(j) = 0$. Show that

$$-\frac{\zeta'(s)}{\zeta(s)} = \sum_{j=1}^{\infty} \frac{\Lambda(j)}{j^s}.$$

3. The Chebyshev functions ϑ and ψ are defined by

$$\vartheta(x) = \sum_{p \le x} \log p$$

and

$$\psi(x) = \sum_{p^m \le x} \log p,$$

where $p \in \mathcal{P}$ and $x > 0$. The summation in the definition of ψ extends over all combinations of primes p and natural numbers m such that $p^m \le x$.

(a) Show that

$$\psi(x) = \sum_{p \le x} \left\lfloor \frac{\log x}{\log p} \right\rfloor$$

and

$$\log p = \sum_{j \le x} \Lambda(j),$$

where Λ is the von Mangoldt function.

(b) Group the terms of $\psi(x)$ for which m has the same value to get

$$\psi(x) = \vartheta(x) + \vartheta(x^{1/2}) + \vartheta(x^{1/3}) + \cdots.$$

Explain why this sum always has a finite number of terms.

(c) Establish the inequality

$$\vartheta(x) \le \psi(x) \le \pi(x)\log x.$$

(d) If $0 < \alpha < 1$ and $x > 1$, show that

$$\vartheta(x) \ge \sum_{x^\alpha < p \le x} \log p \ge \left(\pi(x) - \pi(x^\alpha)\right)\log x^\alpha.$$

(e) Use the inequality in part (d) and the inequality $\pi(x^\alpha) < x^\alpha$ to establish that

$$\frac{\vartheta(x)}{x} > \alpha\left(\frac{\pi(x)\log x}{x} - \frac{\log x}{x^{1-\alpha}}\right).$$

4. Use question 3 to show that the relations

$$\pi(x) \sim \frac{x}{\log x}, \qquad \vartheta(x) \sim x, \qquad \psi(x) \sim x$$

as $x \to \infty$ are equivalent. The proof of the prime number theorem makes use of the Chebyshev functions as they arise naturally in the analysis.

4.2 Partition Functions

In this section we look at a class of identities that involve products of the form

$$\prod_{j \in \mathcal{A}} \frac{1}{1 - x^j} \tag{4.2.1}$$

or

$$\prod_{j \in \mathcal{A}} (1 + x^j), \tag{4.2.2}$$

where $\mathcal{A} \subseteq \mathbb{N}$ and $|x| < 1$.

Lemma 4.2.1 *The product $\prod_{j=1}^{\infty}(1 + z^j)$ defines a function f that is analytic in $D(0; 1)$. Moreover $1/f$ is also analytic in $D(0; 1)$.*

Proof The geometric series $\sum_{j=0}^{\infty} |z|^j$ converges uniformly to a bounded function in any compact subset of $D(0; 1)$. Theorems 2.2.8 and 2.4.4 therefore imply that the product $\prod_{j=1}^{\infty} (1+z^j)$ is absolutely and uniformly convergent in any compact subset of $D(0; 1)$. Theorem 2.4.1 consequently shows that f is analytic in $D(0; 1)$. For all $z \in D(0; 1)$ and all $n \in \mathbb{N}$ we have $1 + z^n \neq 0$. Thus Theorem 2.4.4 also implies that $f(z) \neq 0$ for all $z \in D(0; 1)$. We infer that $1/f$ is analytic in $D(0; 1)$.   □

The comparison test can be used to demonstrate that $\sum_{j \in A} |z|^j$ converges uniformly for any subset A of $\mathbb{N}$ and the lemma above holds for $\prod_{j \in A} (1 + z^j)$. Taylor's theorem therefore shows that the functions defined by these products have Maclaurin series representations with a unit radius of convergence. Specializing again to a real variable x, we obtain the following result.

Corollary 4.2.2 *For any $A \subseteq \mathbb{N}$ there exist sequences $\{a_j\}$ and $\{b_j\}$ such that*

$$\prod_{j \in A} \frac{1}{1 - x^j} = \sum_{k=0}^{\infty} a_k x^k$$

and

$$\prod_{j \in A} (1 + x^j) = \sum_{k=0}^{\infty} b_k x^k$$

for all $x \in (-1, 1)$.

Theorem 4.2.3 *For all x such that $|x| < 1$,*

$$\prod_{j=0}^{\infty} \left(1 + x^{2^j}\right) = \sum_{j=0}^{\infty} x^j. \tag{4.2.3}$$

Proof It is clear that the product and the series converge absolutely, as $|x| < 1$. Let $\{P_n\}$ and $\{S_n\}$ denote the sequences of partial products and sums respectively. Let P and S denote the limits of $\{P_n\}$ and $\{S_n\}$ respectively.

The first few partial products are

$$P_0(x) = 1 + x,$$
$$P_1(x) = (1 + x)(1 + x^2) = 1 + x + x^2 + x^3,$$
$$P_2(x) = P_1(x)(1 + x^4) = P_1(x) + x^4 + x^5 + x^6 + x^7,$$

and we conjecture that

$$P_n(x) = \sum_{j=0}^{2^{n+1}-1} x^j. \tag{4.2.4}$$

Certainly Eq. (4.2.4) is true for $n = 0$. We use induction to show that it is true for all n.

Suppose Eq. (4.2.4) is true for some $n \geq 0$. Then,

$$P_{n+1}(x) = P_n(x)\left(1 + x^{2^{n+1}}\right)$$

$$= \sum_{j=0}^{2^{n+1}-1} x^j + x^{2^{n+1}} \sum_{j=0}^{2^{n+1}-1} x^j$$

$$= \sum_{j=0}^{2^{n+1}-1} x^j + \sum_{j=0}^{2^{n+1}-1} x^{j+2^{n+1}}$$

$$= \sum_{j=0}^{2^{n+1}-1} x^j + \sum_{j=2^{n+1}}^{2\cdot 2^{n+1}-1} x^j$$

$$= \sum_{j=0}^{2^{n+2}-1} x^j.$$

Equation (4.2.4) thus follows by induction. Evidently,

$$S_{2^{n+1}-1}(x) = P_n(x). \tag{4.2.5}$$

For $|x| < 1$, $S_j \to S$ as $j \to \infty$; therefore, $S_{2^{n+1}-1} \to S$ as $n \to \infty$. Taking limits of both sides of Eq. (4.2.5) thus gives

$$S(x) = P(x)$$

whenever $|x| < 1$. $\qquad\qquad\Box$

Although the preceding proof is straightforward, it may leave the reader with a certain empty feeling about the approach. It is one matter to be given an identity to prove and quite a different matter to discover such an identity. How did mathematicians come up with equations such as (4.2.3)? This question leads to a high mathematical plateau where analysis, number theory and combinatorics meet. Let us revisit the product and see how the identity could be discovered.

We know that the product in Eq. (4.2.3) must have a Maclaurin series representation for $|x| < 1$. Thus, there exists a sequence $\{a_n\}$ such that

$$\prod_{j=0}^{\infty}\left(1+x^{2^j}\right) = \sum_{j=0}^{\infty} a_j x^j.$$

Let P_n be as defined in the proof above. Any exponent of x in the expansion of P_n must be of the form

$$\sum_{j=0}^{n} \alpha_j 2^j,$$

where α_j is either 0 or 1. The exponents of x in the power series must therefore be of a similar form. For any $m \in \mathbb{N}$ the coefficient a_m corresponds to the number of distinct ways that m can be represented in the form

$$m = \sum_{j=0}^{\mu} \beta_j 2^j, \tag{4.2.6}$$

where $\mu = \lfloor \log m / \log 2 \rfloor$ and β_j is either 0 or 1. For instance, if $a_m = 0$, then m cannot be written in the form above; if $a_m = 2$, then there exist two sets of coefficients $\{\beta_0, \beta_1, \ldots, \beta_\mu\}$ and $\{\gamma_0, \gamma_1, \ldots, \gamma_\mu\}$ such that

$$m = \sum_{j=0}^{\mu} \beta_j 2^j = \sum_{j=0}^{\mu} \gamma_j 2^j, \tag{4.2.7}$$

and $\beta_j \neq \gamma_j$ for at least one $j \in \{0, 1, \ldots, \mu\}$.

It can be shown that every natural number has a unique binary representation. In other words, every $m \in \mathbb{N}$ can be written in the form (4.2.6) and, if Eq. (4.2.7) is true, then $\beta_j = \gamma_j$ for all $j \in \{0, 1, \ldots, \mu\}$. In this manner we deduce that $a_m = 1$ for all $m \in \mathbb{N}$. Since $P(0) = 1$, we know that $a_0 = 1$; hence we arrive at Eq. (4.2.3).

These arguments bring to the fore the connection between power series coefficients and the representation of a natural number as a sum of natural numbers from a given set. Given a subset $\mathcal{A} \subseteq \mathbb{N}$ and an number $n \in \mathbb{N}$, the essence of the problem is to find the number $A(n)$ of ways that n can be expressed as a sum of numbers in $\mathcal{A}$. Each such representation of n is called a **partition** of n, the summands are called **parts** and A is called a **partition function**. The infinite product that yields the values of the partition function as coefficients in the Maclaurin series is sometimes called the **generating function**. For the example above,

$$\mathcal{A} = \{2^k : k \in \mathbb{N} \cup \{0\}\},$$

and the partition problem entails finding the number of ways that a given n can be expressed as a sum of elements in $\mathcal{A}$. Partition functions do not take into account the order of the parts in the sum. For instance, $2^0 + 2^1$ and $2^1 + 2^0$ are considered the same partition of 3. The partition problem at hand does not allow elements in $\mathcal{A}$ to be repeated. Thus, $2^1 + 2^2$ is a partition of 6, but $2^1 + 2^1 + 2^1$ is not considered a partition of 6.

The nature of the partition problem is linked to the form of the product. Products of the form (4.2.2) lead to partition problems where the parts are distinct. For example,

$$\prod_{j=1}^{\infty} \left(1 + x^{j^2}\right) = 1 + \sum_{j=1}^{\infty} A(j)x^j,$$

where $A(j)$ is the number of partitions of j into distinct parts which are squares. For instance, $A(1) = 1$, $A(2) = A(3) = 0$ and $A(4) = 1$. Since $25 = 5^2 = 3^2 + 4^2$, we see that $A(25) = 2$. Notice that the reasoning we have used depends only on algebraic properties of the product and the series, and does not require the convergence of either.

Example 4.2.1 Consider the product

$$P(x) = \prod_{j=1}^{\infty} \left(1 + x^{2j}\right).$$

The first few partial products are

$$P_1(x) = 1 + x^2,$$
$$P_2(x) = 1 + x^2 + x^4 + x^6,$$
$$P_3(x) = 1 + x^2 + x^4 + 2x^6 + x^8 + x^{10} + x^{12},$$

and it is clear that the Maclaurin series for P is of the form

$$P(x) = 1 + \sum_{j=1}^{\infty} A(2j)x^{2j}.$$

The coefficient of x^{2j} is the number of partitions of $2j$ with distinct positive even parts. The first seven such coefficients are 1, 1, 2, 2, 3, 4, 5.   △

Example 4.2.1 illustrates a **restricted partition problem** because the elements of $\mathcal{A}$ cannot be repeated in a partition, and $\mathcal{A} \neq \mathbb{N}$. The other extreme is the **unrestricted partition problem**, which consists of finding the number of ways that a given positive integer can be expressed as the sum of natural numbers. For this problem $\mathcal{A} = \mathbb{N}$, and repetitions are allowed. Following the standard notation, we write the partition function for this problem as p. For example,

$$5 = 1+1+1+1+1 = 2+1+1+1 = 3+1+1 = 4+1 = 2+2+1 = 2+3$$

and consequently $p(5) = 7$.

Theorem 4.2.4 (Euler) *If $|x| < 1$, then*

$$\prod_{j=1}^{\infty} \frac{1}{1-x^j} = 1 + \sum_{j=1}^{\infty} p(j)x^j.$$

Proof Since $|x| < 1$, we can write

$$\prod_{j=1}^{\infty} \frac{1}{1-x^j} = \prod_{j=1}^{\infty} \sum_{k=0}^{\infty} x^{jk}$$

$$= \prod_{j=1}^{\infty} (1 + x^j + x^{2j} + \cdots).$$

For any $m \in \mathbb{N}$ we obtain a term x^m in the product on the right hand side by taking non-negative integers $k_1, k_2, \ldots, k_m$ such that

$$m = \sum_{j=1}^{m} jk_j = k_1 + 2k_2 + \cdots + mk_m.$$

This sum is a partition of m with k_j parts equal to j for each $j \in \{1, 2, \ldots, m\}$. The result follows immediately.   □

The partition function p grows exponentially with n. For example,

$$p(5) = 7,$$
$$p(10) = 42,$$
$$p(50) = 204, 226,$$
$$p(100) = 190, 569, 292.$$

The generating function for p can be used to get a rough upper bound. First, however, we derive an upper bound for

$$P(x) = \prod_{j=1}^{\infty} \frac{1}{1-x^j}.$$

Lemma 4.2.5 *For all $x \in (0, 1)$,*

$$\log P(x) < \frac{\pi^2 x}{6(1 - x)}.$$

Proof We have

$$\log P(x) = -\sum_{k=1}^{\infty} \log(1 - x^k)$$

$$= \sum_{k=1}^{\infty} \sum_{j=1}^{\infty} \frac{x^{jk}}{j},$$

where we have used the Maclaurin series for $\log(1 - x^k)$. Theorem 2.2.5 shows that $\sum_{k=1}^{\infty} \log(1 - x^k)$ is absolutely convergent since $\sum_{k=1}^{\infty} x^k$ is absolutely convergent. Theorem 1.6.8 thus implies

$$\log P(x) = \sum_{j=1}^{\infty} \frac{1}{j} \sum_{k=1}^{\infty} x^{jk}$$

$$= \sum_{j=1}^{\infty} \frac{1}{j} \left(\frac{1}{1 - x^j} - 1 \right)$$

$$= \sum_{j=1}^{\infty} \frac{x^j}{j(1 - x^j)}. \qquad (4.2.8)$$

Since

$$\frac{1 - x^j}{1 - x} = \sum_{k=0}^{j-1} x^k \geq \sum_{k=0}^{j-1} x^{j-1} = jx^{j-1} \qquad (4.2.9)$$

for each $x \in (0, 1)$, we deduce that

$$\frac{j(1 - x)}{x} \leq \frac{1 - x^j}{x^j},$$

and so

$$\frac{x^j}{j(1 - x^j)} \leq \frac{x}{j^2(1 - x)}. \qquad (4.2.10)$$

As inequality (4.2.9) is strict for all $j > 1$, Eq. (4.2.8) and inequality (4.2.10) imply

$$\log P(x) < \frac{x}{1-x} \sum_{j=1}^{\infty} \frac{1}{j^2},$$

and the result follows because $\zeta(2) = \pi^2/6$. $\qquad\qquad\qquad\qquad\qquad\qquad\qquad\qquad$ □

Theorem 4.2.6 *For all $n \in \mathbb{N}$,*

$$p(n) < e^{\pi\sqrt{2n/3}}. \qquad\qquad (4.2.11)$$

Proof For each $x \in (0, 1)$ and $n \in \mathbb{N}$, Theorem 4.2.4 implies

$$P(x) > p(n)x^n;$$

consequently,

$$\log P(x) > \log p(n) + n \log x.$$

Lemma 4.2.5 thus gives

$$\log p(n) < \frac{\pi^2 x}{6(1-x)} + n \log \frac{1}{x}. \qquad\qquad (4.2.12)$$

For all $w \neq 0$,

$$\log(1 + w) < w$$

(since $e^w > 1 + w$), and therefore

$$\log \frac{1}{x} = \log\left(1 + \left(\frac{1}{x} - 1\right)\right) < \frac{1}{x} - 1 = \frac{1-x}{x}.$$

Let

$$\xi(x) = \frac{x}{1-x} \geq 0$$

for all $x \in [0, 1)$. Since $\xi(0) = 0$ and

$$\lim_{x \to 1^-} \frac{x}{1-x} = \infty,$$

the continuity of the function ξ shows that its range is $[0, \infty)$. Writing ξ instead of $\xi(x)$ for convenience, we see that inequality (4.2.12) implies

$$\log p(n) < \frac{\pi^2}{6}\xi + \frac{n}{\xi} \qquad (4.2.13)$$

whenever $\xi > 0$. If

$$g(\xi) = \frac{\pi^2}{6}\xi + \frac{n}{\xi},$$

then

$$g'(\xi) = \frac{\pi^2}{6} - \frac{n}{\xi^2} = \frac{\xi^2\pi^2 - 6n}{6\xi^2}.$$

Thus $g'(\xi) = 0$ if $\xi = \sqrt{6n}/\pi$. Since $g''(\xi) = 2n/\xi^3 > 0$, g' is increasing at all $\xi > 0$. We therefore conclude that the function g has a minimum at

$$\xi = \frac{\sqrt{6n}}{\pi}.$$

Using this value in inequality (4.2.13) gives

$$\log p(n) < \frac{\pi\sqrt{n}}{\sqrt{6}} + \frac{\pi\sqrt{n}}{\sqrt{6}} = \pi\sqrt{\frac{2n}{3}},$$

which implies inequality (4.2.11).   □

The proofs of the last two results follow those given by Apostol [8], who shows that with a small modification the proofs lead to an upper bound of the form

$$p(n) < \frac{\pi e^{K\sqrt{n}}}{\sqrt{6(n-1)}},$$

where $K = \pi\sqrt{2/3}$. It is remarkable that there is a formula for $p(n)$. The formula is due to Hardy and Ramanujan, and it was put into its final form by Rademacher. The reader can find the formula for $p(n)$ along with a proof in [5]. One outcome of Rademacher's work is the asymptotic relation

$$p(n) \sim \frac{e^{K\sqrt{n}}}{4n\sqrt{3}}.$$

The proof of Theorem 4.2.4 gives us an insight into the nature of partition problems arising from generating functions of the form (4.2.1). Each factor of these products can be written as a geometric series and this observation means that such functions lead to partition problems where repeated parts are allowed. For example,

$$\prod_{j=1}^{\infty} \frac{1}{1-x^{j^2}} = 1 + \sum_{j=1}^{\infty} B(j)x^j,$$

where $B(j)$ is the number of partitions of j into parts which are squares, with repetitions allowed. For this case, $B(1) = B(2) = B(3) = 1$ and $B(4) = 2$, since $4 = 1^2 + 1^2 + 1^2 + 1^2 = 2^2$.

Example 4.2.2 We can use generating functions to show that the number of partitions of a number $k \in \mathbb{N}$ into distinct parts is equal to the number of partitions of k into odd parts. The generating function for the former partition problem is

$$F(x) = \prod_{j=1}^{\infty} \left(1 + x^j\right),$$

for all x such that $|x| < 1$. The partition function for the latter case allows repetitions and therefore the generating function is

$$G(x) = \prod_{j=1}^{\infty} \frac{1}{1 - x^{2j-1}},$$

where $|x| < 1$. The functions F and G must have Maclaurin series, and it suffices to show that $F(x) = G(x)$ whenever $|x| < 1$, since the uniqueness of the Maclaurin series implies that the partition functions must be the same. Since

$$(1 + x^j)(1 - x^j) = 1 - x^{2j},$$

we have

$$F(x) = \prod_{j=1}^{\infty}(1 + x^j)$$

$$= \prod_{j=1}^{\infty} \frac{1 - x^{2j}}{1 - x^j}$$

$$= \prod_{j=1}^{\infty} \left(1 - x^{2j}\right) \prod_{j=1}^{\infty} \frac{1}{(1 - x^{2j-1})(1 - x^{2j})}$$

$$= \prod_{j=1}^{\infty} \frac{1}{1 - x^{2j-1}}$$

$$= G(x),$$

as required.   △

The product

$$\prod_{j=1}^{\infty}\left(1-x^j\right) \tag{4.2.14}$$

is not of the form (4.2.2) owing to the minus signs in the factors. There is nonetheless a remarkable identity between this product and a power series that has nonzero terms only when the exponent is of the form

$$g(k) = \frac{3k^2 - k}{2}$$

or

$$g(-k) = \frac{3k^2 + k}{2},$$

where $k \in \mathbb{N}$. The numbers $g(k)$ and $g(-k)$ are called **pentagonal numbers**. The first few pentagonal numbers are 1,2,5,7,12,15. Euler proved that

$$\prod_{j=1}^{\infty}\left(1-x^j\right) = 1 - x - x^2 + x^5 + x^7 - x^{12} - x^{15} + \cdots$$

$$= 1 + \sum_{j=1}^{\infty}(-1)^j\left(x^{g(j)} + x^{g(-j)}\right)$$

whenever $|x| < 1$. The result is known as the Euler pentagonal number theorem. We prove it in the next section. At present, we interpret the product in terms of partitions.

Note that

$$\prod_{j=1}^{n}(1-x^j) = 1 - x - \sum_{k=2}^{n}\left(x^k\prod_{j=1}^{k-1}(1-x^j)\right)$$

for all $n \in \mathbb{N}$, by induction: this equation holds for $n = 1$, and if it holds for a particular $n \in \mathbb{N}$ then

$$1 - x - \sum_{k=2}^{n+1}\left(x^k\prod_{j=1}^{k-1}(1-x^j)\right)$$

$$= 1 - x - \sum_{k=2}^{n} \left(x^k \prod_{j=1}^{k-1} (1 - x^j) \right) - x^{n+1} \prod_{j=1}^{n} (1 - x^j)$$

$$= \prod_{j=1}^{n} (1 - x^j) - x^{n+1} \prod_{j=1}^{n} (1 - x^j)$$

$$= \prod_{j=1}^{n} (1 - x^j)(1 - x^{n+1})$$

$$= \prod_{j=1}^{n+1} (1 - x^j).$$

Taking limits as $n \to \infty$, we obtain a Maclaurin series for $\prod_{j=1}^{\infty}(1 - x^j)$. Let

$$\prod_{j=1}^{\infty} \left(1 - x^j \right) = 1 + \sum_{j=1}^{\infty} a_j x^j.$$

The coefficients a_j are determined by the partitions of j, but a_j is not simply counting the partitions of j into distinct parts. The minus signs in the factors indicate that the contribution of a partition to the coefficient depends on whether the partition contains an even or an odd number of parts. Consider, for example, the terms in the product that can produce an x^m term in the series. An x^m term arises from a partition of m into distinct parts. Let

$$m = m_1 + m_2 + \cdots + m_N$$

be a partition of m into N distinct parts. Then the corresponding x^m term is

$$(-1)^N x^m.$$

If N is even then the contribution is x^m; if N is odd then the contribution is $-x^m$. We thus have the following result.

Theorem 4.2.7 *If $|x| < 1$, then*

$$\prod_{j=1}^{\infty} \left(1 - x^j \right) = 1 + \sum_{j=1}^{\infty} (A_e(j) - A_o(j)) \, x^j,$$

where $A_e(j)$ and $A_o(j)$ denote, respectively, the numbers of partitions of j into even and odd numbers of distinct parts.

4.3 The Jacobi Triple Product Identity

The Jacobi triple product identity arises from the theory of theta functions. It is fundamental to much of the work on partition problems and it can be used to establish the Euler pentagonal number theorem along with related results. The proof we give for this identity follows that given by Andrews [6]. Different proofs can be found in [8, 30] and [21]. Andrews's proof has the advantage that we need not appeal to any complex analysis (in particular, properties of Laurent series) and the proof brings to the fore two identities by Euler that are of interest in their own right.

The two identities by Euler that are needed in the proof are special cases of the relation

$$\prod_{j=0}^{\infty} \frac{1 + \alpha x q^j}{1 - \beta x q^j} = 1 + \sum_{j=1}^{\infty} \prod_{k=1}^{j} \left(\frac{\beta + \alpha q^{k-1}}{1 - q^k} \right) x^j,$$

where α, β, x and q are numbers such that $\beta \neq 0$, $|\beta x| < 1$ and $|q| < 1$. To prove this relation we first establish it for the special case when $\beta = 1$. Let

$$H(x) = \prod_{j=0}^{\infty} \frac{1 + \alpha x q^j}{1 - x q^j},$$

where $|q| < 1$ and $|x| < 1$. Since $q^n \to 0$ as $n \to \infty$, there exists N such that $\alpha x q^j > -1$ for all $j \geq N$. Therefore Theorem 2.5.1 shows that $\prod_{j=0}^{\infty} (1 + \alpha x q^j)$ and $\prod_{j=0}^{\infty} (1 - x q^j)$ are absolutely convergent and uniformly convergent with respect to x in any interval: since

$$\sum_{j=0}^{\infty} |\alpha x q^j| \leq \alpha \sum_{j=0}^{\infty} |q|^j,$$

the series $\sum_{j=0}^{\infty} |\alpha x q^j|$ is uniformly convergent to a bounded function, by the Weierstrass M-test. The function H is thus well defined and continuous whenever $|q| < 1$ and $|x| < 1$.

The definition of H implies that

$$H(qx) = \prod_{j=0}^{\infty} \frac{1 + \alpha x q^{j+1}}{1 - x q^{j+1}}.$$

If $\alpha x q^j = -1$ for some $j > 0$, then $H(x) = H(qx) = 0$. In the remaining case we have

$$\frac{H(x)}{H(qx)} = \frac{\frac{1+\alpha x}{1-x}}{\frac{1+\alpha xq}{1-xq}} \cdot \frac{\frac{1+\alpha xq}{1-xq}}{\frac{1+\alpha xq^2}{1-xq^2}} \cdots$$

$$= \frac{1+\alpha x}{1-x}.$$

In any case, H is a solution to the functional equation

$$(1-x)\Phi(x) = (1+\alpha x)\Phi(qx). \tag{4.3.1}$$

We will use this equation to deduce the power series expansion for H. The idea is to show that there is only one continuous solution to Eq. (4.3.1) that satisfies

$$\Phi(0) = 1. \tag{4.3.2}$$

We then develop formally a Maclaurin series for Φ and show that this is also a solution to Eq. (4.3.1) that satisfies equation (4.3.2).

Lemma 4.3.1 *There is only one solution to Eq. (4.3.1) such that Φ is continuous and satisfies condition (4.3.2).*

Proof Suppose that there exist two distinct solutions Φ_1 and Φ_2 to Eq. (4.3.1) that are continuous at 0 and satisfy equation (4.3.2). Let $\Delta = \Phi_1 - \Phi_2$. Then Δ also satisfies equation (4.3.1) and

$$\Delta(0) = 0. \tag{4.3.3}$$

Since Φ_1 and Φ_2 are distinct, there is an $\hat{x}$ such that $0 < |\hat{x}| < 1$ and

$$\Delta(\hat{x}) \neq 0. \tag{4.3.4}$$

As $|q\hat{x}| < 1$, Eq. (4.3.1) implies

$$\Delta(\hat{x}) = \frac{1+\alpha\hat{x}}{1-\hat{x}}\Delta(q\hat{x})$$

$$= \frac{1+\alpha\hat{x}}{1-\hat{x}} \cdot \frac{1+\alpha\hat{x}q}{1-\hat{x}q}\Delta(q^2\hat{x}).$$

By induction it follows that

$$\Delta(\hat{x}) = \prod_{j=0}^{n-1}\left(\frac{1+\alpha\hat{x}q^j}{1-\hat{x}q^j}\right)\Delta(q^n\hat{x})$$

for any $n \in \mathbb{N}$; consequently,

$$\Delta(\hat{x}) = \prod_{j=0}^{\infty} \left(\frac{1 + \alpha \hat{x} q^j}{1 - \hat{x} q^j} \right) \lim_{n \to \infty} \Delta(q^n \hat{x}). \tag{4.3.5}$$

The function Δ is continuous at $0 = \lim_{n \to \infty} q^n \hat{x}$ and so

$$\lim_{n \to \infty} \Delta(q^n \hat{x}) = \Delta(0) = 0.$$

Equation (4.3.5) therefore shows that

$$\Delta(\hat{x}) = 0,$$

in contradiction to Eq. (4.3.4).   □

Suppose that there exists a solution to Eq. (4.3.1) of the form

$$\Phi(x) = \sum_{j=0}^{\infty} c_j x^j, \tag{4.3.6}$$

where $|x| < 1$, that satisfies condition (4.3.2). We proceed to find the coefficients in this Maclaurin series. First, Eq. (4.3.2) implies that $c_0 = 1$. We have

$$(1 - x)\Phi(x) = (1 - x) \sum_{j=0}^{\infty} c_j x^j$$

$$= \sum_{j=0}^{\infty} c_j x^j - \sum_{j=1}^{\infty} c_{j-1} x^j$$

$$= 1 + \sum_{j=1}^{\infty} \left(c_j - c_{j-1} \right) x^j,$$

and similarly

$$(1 + \alpha x)\Phi(qx) = (1 + \alpha x) \sum_{j=0}^{\infty} c_j q^j x^j$$

$$= 1 + \sum_{j=1}^{\infty} \left(c_j q^j + \alpha c_{j-1} q^{j-1} \right) x^j.$$

Equation (4.3.1) thus gives

$$\sum_{j=1}^{\infty} \left(c_j - c_{j-1} \right) x^j = \sum_{j=1}^{\infty} \left(c_j q^j + \alpha c_{j-1} q^{j-1} \right) x^j,$$

and the uniqueness of the Maclaurin expansion implies

$$c_j - c_{j-1} = c_j q^j + \alpha c_{j-1} q^{j-1},$$

so that

$$c_j = \frac{1 + \alpha q^{j-1}}{1 - q^j} c_{j-1} \tag{4.3.7}$$

for each $j \in \mathbb{N}$. This recursive relation can be applied j times to get

$$c_j = \prod_{k=1}^{j} \frac{1 + \alpha q^{k-1}}{1 - q^k}.$$

Let us now investigate the convergence of this Maclaurin series. If $q = 0$ then $c_j = 1 + \alpha$ for all $j > 0$, and the series (4.3.6) converges since $|x| < 1$. Suppose therefore that $q \neq 0$. If $\alpha = -1/q^{m-1}$ for some $m \in \mathbb{N}$, then $c_j = 0$ for all $j \geq m$, and the series (4.3.6) reduces to a polynomial. Suppose that there is no $m \in \mathbb{N}$ such that $\alpha = -1/q^{m-1}$. Then $c_j \neq 0$ for all $j \in \mathbb{N}$, and if $x \neq 0$ then

$$\lim_{j \to \infty} \left| \frac{c_{j+1} x^{j+1}}{c_j x^j} \right| = \lim_{j \to \infty} \left| \frac{1 + \alpha q^j}{1 - q^{j+1}} \right| |x| = |x| < 1.$$

The ratio test thus implies that the series (4.3.6) has a unit radius of convergence. In any event, the series (4.3.6) converges for all α and all x such that $|x| < 1$. Moreover, the power series defining Φ must be continuous whenever $|x| < 1$. By construction the series satisfies equations (4.3.1) and (4.3.2). The product H, however, is also a solution. Lemma 4.3.1 therefore yields the following result.

Theorem 4.3.2 *For all α, q and x such that $|q| < 1$ and $|x| < 1$,*

$$\prod_{j=0}^{\infty} \frac{1 + \alpha x q^j}{1 - x q^j} = 1 + \sum_{j=1}^{\infty} \prod_{k=1}^{j} \left(\frac{1 + \alpha q^{k-1}}{1 - q^k} \right) x^j. \tag{4.3.8}$$

Corollary 4.3.3 *For all α, β, q and x such that $|q| < 1$, $|\beta x| < 1$ and $\beta \neq 0$,*

$$\prod_{j=0}^{\infty} \frac{1 + \alpha x q^j}{1 - \beta x q^j} = 1 + \sum_{j=1}^{\infty} \prod_{k=1}^{j} \left(\frac{\beta + \alpha q^{k-1}}{1 - q^k} \right) x^j. \tag{4.3.9}$$

Proof Since $|\beta x| < 1$ and $\beta \neq 0$, we can replace x with βx and α with α/β in Eq. (4.3.8). These substitutions give

$$\prod_{j=0}^{\infty} \frac{1 + \alpha x q^j}{1 - \beta x q^j} = 1 + \sum_{j=1}^{\infty} \prod_{k=1}^{j} \left(\frac{1 + \frac{\alpha}{\beta} q^{k-1}}{1 - q^k} \right) \beta^j x^j$$

$$= 1 + \sum_{j=1}^{\infty} \prod_{k=1}^{j} \left(\frac{\beta + \alpha q^{k-1}}{1 - q^k} \right) x^j.$$

□

Corollary 4.3.4 (Euler) *If $|q| < 1$, then*

$$\prod_{j=0}^{\infty} \left(1 + x q^j \right) = 1 + \sum_{j=1}^{\infty} \frac{q^{\frac{j(j-1)}{2}} x^j}{\prod_{k=1}^{j} (1 - q^k)} \tag{4.3.10}$$

for all x, and

$$\prod_{j=0}^{\infty} \frac{1}{1 - x q^j} = 1 + \sum_{j=1}^{\infty} \frac{x^j}{\prod_{k=1}^{j} (1 - q^k)} \tag{4.3.11}$$

whenever $|x| < 1$.

Proof To prove Eq. (4.3.10), note first that both sides are equal to 1 if $x = 0$. Let $\alpha = 1$ in Eq. (4.3.9) and fix $x \neq 0$.

We first show that

$$\prod_{j=0}^{\infty} \frac{1 + x q^j}{1 - \beta x q^j}$$

is uniformly convergent with respect to β on any closed interval I included in $(-\tau, \tau)$, where $\tau = \min\{1, 1/|x|\}$. Thus $|\beta| < 1$ and $|\beta x| < 1$ for all $\beta \in I$. Note that

$$\frac{1 + x q^j}{1 - \beta x q^j} - 1 = \frac{(1 + \beta) x q^j}{1 - \beta x q^j} \to 0 \tag{4.3.12}$$

as $j \to \infty$. Moreover the right hand side of Eq. (4.3.12) is a function of β that is continuous, and hence bounded, on I. Thus by Theorem 2.5.1 it suffices to show that

$$\sum_{j=0}^{\infty} \left| \frac{(1 + \beta) x q^j}{1 - \beta x q^j} \right|$$

converges uniformly to a bounded function. For this purpose we may assume that $q \neq 0$, the desired result being obvious otherwise. Since τ has been chosen so that $\beta \neq -1$ and $|\beta x| < 1$, we have

$$\left| \frac{(1+\beta)xq^{n+1}}{1-\beta xq^{n+1}} \cdot \frac{1-\beta xq^n}{(1+\beta)xq^n} \right| = \left| \frac{q(1-\beta xq^n)}{1-\beta xq^{n+1}} \right| \to |q| < 1$$

as $n \to \infty$. Consequently the series does converge uniformly, by Theorem 1.4.23, to a continuous and hence bounded function, by Corollary 1.4.18. The product therefore converges to a function continuous on a neighbourhood of 0, by Theorem 2.5.2(1). It converges to the constant function 0 if $xq^j = -1$ for some non-negative integer j.

We also confirm the uniform convergence of

$$\sum_{j=1}^{\infty} \prod_{k=1}^{j} \left(\frac{\beta + q^{k-1}}{1-q^k} \right) x^j = \sum_{j=1}^{\infty} \prod_{k=1}^{j} \frac{\beta x + q^{k-1}x}{1-q^k}$$

with respect to β on any closed interval included in $(-\tau, \tau)$. The product is a function of β that is continuous, and therefore bounded, on any closed interval. If $\beta = -q^l$ for some $l \in \mathbb{N} \cup \{0\}$, then the series is a finite sum, and so we assume that $\beta + q^{k-1} \neq 0$ for all k. Since $x \neq 0$ it follows that

$$\left| \frac{\prod_{k=1}^{n+1} \frac{\beta x + q^{k-1}x}{1-q^k}}{\prod_{k=1}^{n} \frac{\beta x + q^{k-1}x}{1-q^k}} \right| = \left| \frac{\beta x + q^n x}{1-q^{n+1}} \right| \to |\beta x| < 1$$

as $n \to \infty$. Hence the series converges uniformly, by Theorem 1.4.23, to a function that is continuous, by Corollary 1.4.18.

We can thus take the limit as $\beta \to 0$ on both sides of Eq. (4.3.9) to get

$$\prod_{j=0}^{\infty} \left(1 + xq^j \right) = 1 + \sum_{j=1}^{\infty} \prod_{k=1}^{j} \left(\frac{q^{k-1}}{1-q^k} \right) x^j.$$

Equation (4.3.10) follows from the relation

$$\prod_{k=1}^{j} q^{k-1} = q^{1+2+\cdots+(j-1)} = q^{\frac{j(j-1)}{2}}.$$

Equation (4.3.11) follows immediately from (4.3.8) using $\alpha = 0$.   □

The series on the right hand side of Eq. (4.3.11) has a unit radius of convergence. Hence the equation cannot be extended beyond the interval $(-1, 1)$.

Theorem 4.3.5 (Jacobi Triple Product Identity) *If $x \neq 0$ and $|q| < 1$, then*

$$\prod_{j=0}^{\infty} \left(1 - q^{2j+2}\right) \left(1 + xq^{2j+1}\right) \left(1 + x^{-1}q^{2j+1}\right) = \sum_{j=0}^{\infty} q^{j^2} x^j + \sum_{j=1}^{\infty} q^{j^2} x^{-j}.$$

(4.3.13)

Proof For the sake of succinctness we shall adopt the standard notation of writing the right hand side of Eq. (4.3.13) as

$$\sum_{j=-\infty}^{\infty} q^{j^2} x^j.$$

Let

$$R(q) = \prod_{j=0}^{\infty} \left(1 - q^{2j+2}\right)$$

and

$$S(x, q) = \prod_{j=0}^{\infty} \left(1 + xq^{2j+1}\right).$$

Since $|q| < 1$, these products both converge by Corollary 2.2.7, though possibly to 0 in the case of $S(x, q)$.

We suppose first that $|x| > |q|$. We can replace q by q^2 and x by xq in Eq. (4.3.10) to get

$$S(x, q) = 1 + \sum_{j=1}^{\infty} \frac{q^{j^2} x^j}{\prod_{k=1}^{j} \left(1 - q^{2k}\right)}.$$

(4.3.14)

Since

$$R(q) = \prod_{k=0}^{\infty} (1 - q^{2k+2})$$

$$= \prod_{k=0}^{j-1} (1 - q^{2k+2}) \prod_{k=j}^{\infty} (1 - q^{2k+2})$$

$$= \prod_{k=1}^{j} (1 - q^{2k}) \prod_{k=0}^{\infty} (1 - q^{2k+2j+2}),$$

it follows that

$$\frac{1}{R(q)} \prod_{k=0}^{\infty} (1 - q^{2k+2j+2}) = \frac{1}{\prod_{k=1}^{j}(1 - q^{2k})}.$$

Therefore Eq. (4.3.14) implies that

$$S(x, q) = 1 + \frac{1}{R(q)} \sum_{j=1}^{\infty} q^{j^2} x^j \prod_{k=0}^{\infty} \left(1 - q^{2k+2j+2}\right)$$

$$= \frac{1}{R(q)} \sum_{j=0}^{\infty} q^{j^2} x^j \prod_{k=0}^{\infty} \left(1 - q^{2k+2j+2}\right).$$

If j is a negative integer, then $2k + 2j + 2 = 0$ for $k = -j - 1 \geq 0$; consequently,

$$\prod_{k=0}^{\infty} \left(1 - q^{2k+2j+2}\right) = 0.$$

We can thus extend the summation to get

$$R(q)S(x, q) = \sum_{j=-\infty}^{\infty} q^{j^2} x^j \prod_{k=0}^{\infty} \left(1 - q^{2k+2j+2}\right). \tag{4.3.15}$$

We show that this series is absolutely convergent. The series

$$\sum_{j=-\infty}^{\infty} |q^{j^2} x^j| = \sum_{j=0}^{\infty} |q^{j^2} x^j| + \sum_{j=1}^{\infty} |q^{j^2} x^{-j}|$$

converges by the ratio test, since

$$\left| \frac{q^{(n+1)^2} x^{n+1}}{q^{n^2} x^n} \right| = |q^{2n+1} x| \rightarrow 0$$

as $n \rightarrow \infty$, and

$$\left| \frac{q^{(n+1)^2}}{x^{n+1}} \cdot \frac{x^n}{q^{n^2}} \right| = \left| \frac{q^{2n+1}}{x} \right| < q^{2n} \rightarrow 0,$$

because $|x| > |q|$. As

$$q^{2(k+j+1)} \geq 0,$$

it follows that

$$\left| \prod_{k=0}^{\infty} \left(1 - q^{2k+2j+2}\right) \right| \leq 1,$$

and so

$$\sum_{j=-\infty}^{\infty} \left| q^{j^2} x^j \prod_{k=0}^{\infty} \left(1 - q^{2k+2j+2}\right) \right|$$

converges by the comparison test.

If we replace q by q^2 and use $x = -q^{2j+2}$, Eq. (4.3.10) gives (with the index k instead of j)

$$\prod_{k=0}^{\infty} \left(1 - q^{2k+2j+2}\right) = 1 + \sum_{k=1}^{\infty} \frac{(-1)^k q^{k^2+2kj+k}}{\prod_{l=1}^{k}(1 - q^{2l})};$$

consequently,

$$R(q)S(x,q) = \sum_{j=-\infty}^{\infty} q^{j^2} x^j \left(1 + \sum_{k=1}^{\infty} \frac{(-1)^k q^{k^2+2jk+k}}{\prod_{l=1}^{k}(1 - q^{2l})}\right)$$

$$= \sum_{j=-\infty}^{\infty} q^{j^2} x^j + \sum_{j=-\infty}^{\infty} \sum_{k=1}^{\infty} \frac{(-1)^k q^{k^2+2jk+j^2+k} x^j}{\prod_{l=1}^{k}(1 - q^{2l})}$$

$$= \sum_{j=-\infty}^{\infty} q^{j^2} x^j + \sum_{j=-\infty}^{\infty} \sum_{k=1}^{\infty} \frac{(-1)^k x^{-k} q^k q^{(j+k)^2} x^{j+k}}{\prod_{l=1}^{k}(1 - q^{2l})}.$$

The first term on the right hand side is an absolutely convergent series. As the series in Eq. (4.3.15) is also absolutely convergent and a sum of absolutely convergent series is absolutely convergent, the double series in the expression above is absolutely convergent. Therefore the order of summation can be changed. We thus get

$$R(q)S(x,q) = \sum_{j=-\infty}^{\infty} q^{j^2} x^j + \sum_{k=1}^{\infty} \frac{(-1)^k x^{-k} q^k}{\prod_{l=1}^{k}(1 - q^{2l})} \sum_{j=-\infty}^{\infty} q^{(j+k)^2} x^{j+k}. \qquad (4.3.16)$$

Now, both $j + k$ and j assume each integral value exactly once and therefore

$$\sum_{j=-\infty}^{\infty} q^{(j+k)^2} x^{j+k} = \sum_{j=-\infty}^{\infty} q^{j^2} x^j.$$

Equation (4.3.16) thus implies

$$R(q)S(x, q) = \sum_{j=-\infty}^{\infty} q^{j^2} x^j \left(1 + \sum_{k=1}^{\infty} \frac{(-1)^k x^{-k} q^k}{\prod_{l=1}^{k}(1 - q^{2l})} \right). \tag{4.3.17}$$

Since $|q/x| < 1$, we can replace q by q^2 and replace x by $-q/x$ in Eq. (4.3.11). These substitutions yield

$$1 + \sum_{k=1}^{\infty} \frac{(-1)^k x^{-k} q^k}{\prod_{l=1}^{k}(1 - q^{2l})} = \prod_{k=0}^{\infty} \frac{1}{1 + x^{-1} q^{2k+1}} = \frac{1}{S(x^{-1}, q)}.$$

Equation (4.3.17) therefore shows that

$$R(q)S(x, q)S(x^{-1}, q) = \sum_{j=-\infty}^{\infty} q^{j^2} x^j, \tag{4.3.18}$$

which is Eq. (4.3.13), whenever $|x| > |q|$.

Equation (4.3.13) is symmetric with respect to x and x^{-1}. Therefore it holds for x^{-1} if and only if it holds for x. Since $q^2 < 1 = xx^{-1}$, either $|x| > |q|$ or $|x^{-1}| > |q|$; hence, Eq. (4.3.13) is satisfied for all q and x such that $|q| < 1$ and $x \neq 0$.   □

The Jacobi triple product identity (4.3.13) spawns a number of other identities including the Euler pentagonal number theorem. In Eq. (4.3.13), if we replace q with x^m and x with x^n, where $0 < |x| < 1$ and $m, n \in \mathbb{R}$, then

$$\prod_{j=0}^{\infty}(1 - x^{2mj+2m})(1 + x^{2mj+m+n})(1 + x^{2mj+m-n}) = \sum_{j=-\infty}^{\infty} x^{mj^2+jn}. \tag{4.3.19}$$

If we replace q as above but replace x by $-x^n$, then

$$\prod_{j=0}^{\infty}(1 - x^{2mj+2m})(1 - x^{2mj+m+n})(1 - x^{2mj+m-n}) = \sum_{j=-\infty}^{\infty} (-1)^j x^{mj^2+jn}. \tag{4.3.20}$$

Corollary 4.3.6 (Euler Pentagonal Number Theorem) *If* $|x| < 1$, *then*

$$\prod_{j=1}^{\infty}\left(1 - x^j\right) = 1 + \sum_{j=1}^{\infty}(-1)^j \left(x^{g(j)} + x^{g(-j)}\right), \tag{4.3.21}$$

where

$$g(j) = \frac{j(3j-1)}{2}.$$

Proof Both sides are equal to 1 if $x = 0$. If $x \neq 0$, let $m = 3/2$ and $n = 1/2$ in Eq. (4.3.20). Then

$$\prod_{j=0}^{\infty}(1 - x^{3j+3})(1 - x^{3j+2})(1 - x^{3j+1}) = \sum_{j=-\infty}^{\infty} (-1)^j x^{\frac{j(3j+1)}{2}}.$$

Now,

$$\prod_{j=0}^{\infty}(1 - x^{3j+3})(1 - x^{3j+2})(1 - x^{3j+1}) = \prod_{j=1}^{\infty}\left(1 - x^j\right),$$

and

$$\sum_{j=-\infty}^{\infty} (-1)^j x^{\frac{j(3j+1)}{2}} = 1 + \sum_{j=1}^{\infty}(-1)^j x^{\frac{j(3j-1)}{2}} + \sum_{j=1}^{\infty}(-1)^j x^{\frac{j(3j+1)}{2}}$$

$$= 1 + \sum_{j=1}^{\infty}(-1)^j \left(x^{g(j)} + x^{g(-j)}\right),$$

which proves Eq. (4.3.21). $\square$

The next result is an immediate consequence of Corollary 4.3.6 and Theorem 4.2.7.

Corollary 4.3.7 *Let A_e and A_o be the partition functions defined in Theorem 4.2.7. Then*

$$A_e(n) = A_o(n) \tag{4.3.22}$$

for all $n \in \mathbb{N}$, except when

$$n = \frac{j(3j \pm 1)}{2} \tag{4.3.23}$$

for some $j \in \mathbb{N}$. In this case

$$A_e(n) - A_o(n) = (-1)^j.$$

Apostol [8] notes that Euler (c. 1750) proved Corollary 4.3.6 using induction and that later proofs were given by Legendre (c. 1830) and Jacobi (c. 1846). It is of interest to note that there is a purely combinatorial proof of this result by Franklin (c. 1881). This proof can be found in [8] or [30].

The last result we present concerns triangular numbers. Triangular numbers are integers of the form

$$n = \frac{k(k+1)}{2},$$

where k is an integer.

Corollary 4.3.8 (Gauss's Triangular Number Theorem) *If* $|x| < 1$, *then*

$$\sum_{j=0}^{\infty} x^{\frac{j(j+1)}{2}} = \prod_{j=1}^{\infty} \frac{1 - x^{2j}}{1 - x^{2j-1}}. \tag{4.3.24}$$

Proof Both sides are equal to 1 if $x = 0$. If $x \neq 0$, let $m = n = 1/2$ in Eq. (4.3.19). Then,

$$\sum_{j=-\infty}^{\infty} x^{\frac{j(j+1)}{2}} = \prod_{j=0}^{\infty} \left(1 - x^{j+1}\right) \left(1 + x^{j+1}\right) \left(1 + x^{j}\right)$$

$$= 2 \prod_{j=0}^{\infty} \left(1 - x^{2(j+1)}\right) \prod_{j=1}^{\infty} \left(1 + x^{j}\right)$$

$$= 2 \prod_{j=1}^{\infty} \left(1 - x^{2j}\right) \prod_{j=1}^{\infty} \left(1 + x^{j}\right)$$

$$= 2 \prod_{j=1}^{\infty} \frac{1 - x^{2j}}{1 - x^{2j-1}},$$

by Example 4.2.2. But

$$\sum_{j=-\infty}^{\infty} x^{\frac{j(j+1)}{2}} = \sum_{j=0}^{\infty} x^{\frac{j(j+1)}{2}} + \sum_{j=1}^{\infty} x^{\frac{j(j-1)}{2}}$$

$$= 2 \sum_{j=0}^{\infty} x^{\frac{j(j+1)}{2}},$$

and thus we get Eq. (4.3.24).   □

Exercises 4.3

1. Euler's identities in Corollary 4.3.4 can be proved directly in a manner similar to that used to prove Theorem 4.3.2. Let

$$F(x) = \prod_{j=0}^{\infty} \left(1 + xq^j\right)$$

for all $x \in \mathbb{R}$.

(a) Show that F is a solution to

$$\Phi(x) = (1+x)\Phi(qx). \tag{4.3.25}$$

(b) Derive a power series solution to Eq. (4.3.25) and verify that it has an infinite radius of convergence.

(c) Show that there is only one solution to Eq. (4.3.25) that is continuous at 0 and satisfies $\Phi(1) = 1$.

2. The proof of the uniqueness of solutions to equations such (4.3.25) can be avoided if one is prepared to "reverse engineer" the result. Let

$$K(x) = 1 + \sum_{j=1}^{\infty} \frac{x^j}{\prod_{k=1}^{j}(1 - q^k)}$$

for all $x \in \mathbb{R}$.

(a) Show that K is a solution to

$$(1 - x)\Phi(x) = \Phi(qx). \tag{4.3.26}$$

(b) Use the properties of power series (and in particular continuity) to prove Eq. (4.3.11).

3. Show that, whenever $|x| < 1$,

$$\prod_{j=0}^{\infty} \left(1 - x^{2j+1}\right)^2 \left(1 - x^{2j+2}\right) = \sum_{j=-\infty}^{\infty} (-1)^j x^{j^2}$$

and

$$\prod_{j=0}^{\infty} \left(1 + x^{2j+1}\right)^2 \left(1 - x^{2j+2}\right) = \sum_{j=-\infty}^{\infty} x^{j^2}.$$

4. Show that, whenever $|x| < 1$,

$$\prod_{j=0}^{\infty} \left(1 - x^{5j+1}\right) \left(1 - x^{5j+4}\right) \left(1 - x^{5j+5}\right) = \sum_{j=-\infty}^{\infty} (-1)^j x^{\frac{j(5j+3)}{2}}$$

and

$$\prod_{j=0}^{\infty} \left(1 - x^{5j+2}\right) \left(1 - x^{5j+3}\right) \left(1 - x^{5j+5}\right) = \sum_{j=-\infty}^{\infty} (-1)^j x^{\frac{j(5j+1)}{2}}.$$

Chapter 5
Epilogue

At this point we hope that the reader appreciates the important rôle that infinite products play in analysis and other fields of mathematics. Many of the results (and references) in this book have been known by mathematicians for many years. Perhaps most of them date before 1900 bearing a pedigree of luminaries in the field including Euler, Gauss, Cauchy, Dirichlet, Riemann, Jacobi, Weierstrass and Hardy. In the introduction we started with a result by Viète that was published in 1593. One might leave the subject with the feeling that it is a "closed book": what more is there to say about infinite products?

Chapter 4 gives us a glimmer of ongoing work. The Riemann hypothesis is still elusive (at the time of writing) and, as Andrews [5] points out in his preface to the 1998 paperback edition of his book on partitions, the field of partitions has blossomed since 1976. Certainly, there has been a lot of recent work in this area and the related field of q-equations. We make no attempt to summarize this huge field but simply note that this work is impregnated with a prodigious number of relations between infinite products and infinite series, all of which are of interest and some of which are accessible to the non-specialist. Andrews (*loc. cit.*) calls this the "wonderful world of q". A lot of this work is centred around the basic hypergeometric series, and the reader can be brought up to speed with Andrews (*op. cit.*) [21] and Gaspar and Raman [26]. The importance of the ongoing research into infinite products in this field is clear.

The purpose of this short chapter is to give a brief account of some further work on infinite products outside the applications to partitions problems (with one minor exception). Here, it is hoped to put some of the results of the earlier chapters into a wider mathematical context, note some recent research and provide the student with some references for further study. No attempt is made to give a complete or even balanced account of this large field: the choice of what to include is always debatable. We concentrate on results that do not require extensive specialist knowledge to at least appreciate the rôle played by infinite products if not the fine details. We do not prove any of the results but leave the reader to look up the references and, we hope, spark a flame to study these developments.

5.1 Product Representation of Functions

Can we construct an entire function with a given distribution of zeros? The Weierstrass factorization theorem (Theorem 2.9.2) provides an elegant answer to questions such as this. It also provides a representation for a given entire function whose zeros are known. If a bit more information is added (for instance, the order of the function) then we can use the Hadamard factorization theorem [27], which provides sharper information about the entire function g in Theorem 2.9.2. Armed with the Weierstrass factorization theorem we can establish infinite product representations for entire elementary transcendental functions such as $\sin z$ along with $1/\Gamma(z)$. What is the value of knowing (or specifying) the distribution of zeros for entire functions? It turns out that this distribution has a profound impact on the growth of the maximum modulus of a function. Let f be an entire function and

$$M(r) = \sup_{|z|=r} |f(z)|.$$

If, for example, f has the same zeros as $\sin z$, then $M(r)$ cannot grow more slowly than the maximum modulus for $\sin z$ as $r \to \infty$. The relationship is captured in a simple way by Carlson's theorem ([67], p. 186). The zeros of f and growth of $M(r)$ also provide information about deficient values (values f cannot achieve, such as 0 for e^z), and this theory spills over to meromorphic functions via Nevanlinna theory. The reader is directed to the classic work by Boas [12] for a detailed exposition on entire functions and [27] and [62] for accessible accounts of Nevanlinna theory. Zhang [72] provides a modern detailed account of value distributions and asymptotics for entire and meromorphic functions. Another application of the Weierstrass factorization theorem can be found in the theory of Fredholm integral equations. In particular, the eigenvalues of a Fredholm integral equation correspond to the zeros of an entire function D called the Fredholm determinant. A starting point to understand the analytic structure of this determinant is to express D as an infinite product (cf. [31], Chapter 6).

The gamma function certainly finds many applications. A more detailed account of $\Gamma(z)$ including alternative representations can be found in [69]. Identities involving $\Gamma(z)$ and related functions can also be found in [1], where there are also series representations and a complex version of Stirling's approximation that includes higher order terms. The infinite product representations of the gamma function are still at the centre of a number of applications. Recently, for example, new applications have been found in partitions [16] and in products associated with paper folding sequences [3, 4].

Euler's infinite product expansions for $\sin z$ have been known for over 200 years and have proved fruitful in proving expressions such as those for π given by Viète and Wallis. Indeed, these products often form the centrepiece of an exposition on the representation of entire functions. It is thus remarkable that new product representations emerge from this heavily studied field. In 2003, Osler [49] showed that for any positive integers M and N

$$\sin z = z \exp\left(\frac{z}{\pi} \log \frac{M}{N}\right) P(M, N, z), \qquad (5.1.1)$$

where

$$P(M, N, z) = \prod_{n=0}^{\infty} A_n B_n,$$

for

$$A_n = \prod_{j=1}^{M} \left(1 - \frac{z}{\pi(nM + j)}\right)$$

$$B_n = \prod_{j=1}^{N} \left(1 + \frac{z}{\pi(nN + j)}\right).$$

Euler's product corresponds to the special case $M = N = 1$. In fact, as Osler notes, Eq. (5.1.1) appeared on the Cambridge Mathematical Tripos examination in 1904 and Whittaker and Watson ([69], p. 35) give a derivation of it for the special case $M = 2$, $N = 1$. Nearly a hundred years after the examination the proof was published.

Yet another product for $\sin z$ was derived in 2008 by Melnikov [41], *viz.*,

$$\sin z = \frac{2z}{\pi} \prod_{n=1}^{\infty} \left(1 + \frac{4z^2 - \pi^2}{(1 - 4n^2)\pi^2}\right). \qquad (5.1.2)$$

Of particular interest is the method whereby Eq. (5.1.2) is derived. Melnikov uses Green's functions for the two-dimensional Laplace equation. For certain cases, the Green's function is known explicitly. This function can also be determined using the method of images, which yields an infinite product. This approach leads to identities such as (5.1.2) among many others (cf. [42] for a full account).

Aside from the gamma function we have focused mostly on the representation of trigonometric functions. A well-known representation for $\log x$ due to von Seidel is

$$\log x = (x - 1) \prod_{n=1}^{\infty} \left(\frac{2}{1 + x^{1/2^n}} \right), \tag{5.1.3}$$

for $x > 0$. A proof of (5.1.3) can be found in the delightful book by Loya [38], which contains many other interesting results on products. Levin [35] generalized (5.1.3) to

$$\log x = (x - 1) \prod_{n=1}^{\infty} \left(\frac{M}{1 + x^{1/M^n} + x^{2/M^n} + \cdots + x^{(M-1)/M^n}} \right), \tag{5.1.4}$$

where $M \geq 2$ is an integer. Osler [50] established Eq. (5.1.4) using a different approach.

A representation for the square root due to Cantor is

$$\sqrt{\frac{x+1}{x-1}} = \prod_{n=1}^{\infty} \left(1 + \frac{1}{q_n} \right), \tag{5.1.5}$$

where $q_1 = x > 1$ and $q_{n+1} = 2q_n^2 - 1$. (An expression similar to (5.1.5) was derived by Engels, c. 1913.) Fine [22] extended (5.1.5) to kth roots. A feature of Cantor's product is that it converges quickly owing to the square in the definition of q_{n+1}. If we take the first N factors and form the partial product P_N, then the relative error $\epsilon_N = \left| \sqrt{\frac{x+1}{x-1}} - P_N \right|$ satisfies

$$\frac{1}{q_{N+1}} < \epsilon_N < \frac{1}{q_{N+1} - 1},$$

so that the error is approximately $1/q_{N+1}$. For the Cantor product, we see that q_n grows quadratically with n and hence the convergence is called "quadratic". Fine (op. cit.) developed products for roots that converge even faster using a sequence $\{q_n\}$ that grows faster than quadratic. For example, he showed that

$$\sqrt{\frac{x+3}{x-1}} = \prod_{n=1}^{\infty} \left(1 + \frac{1}{q_n} \right) \tag{5.1.6}$$

for $q_1 = x > 1$ and $q_{n+1} = q_n^3 + 3q_n^2 - 3$. Here he shows that the relative error ϵ_N satisfies

$$\frac{2}{q_{N+1}} < \epsilon_N < \frac{2}{q_{N+1} - 1}$$

and that, for instance, if $x \geq 2$ then

$$\epsilon_N < 2 \left(\frac{1}{x}\right)^{3^N}.$$

He notes that for $x = 3$, to approximate $\sqrt{3}$, "the first three factors alone would give an accuracy of 14 decimals, and twelve factors would give well over 300,000 correct decimals." This is seriously fast convergence. We note that other expressions for roots have been derived by Wingler [70] and Nimbran [48].

Finally, before we discuss infinite products of constants, we note that infinite products have been generalized beyond terms that are functions to include terms that are, for example, operators. These products have found applications in approximation theory among other places. The reader is directed to [57], [59] and [60] for more details.

5.2 Infinite Products for Constants

We started this book with Viète's product for the number π and it is fitting that the last section of the epilogue begin with this remarkable product. As noted by Levin [35] this product was "not only the first exact analytic expression ever given for π, but also the first recorded use of an infinite product in mathematics." Viète, a French lawyer and amateur mathematician, published his product in 1593. His derivation is geometrical. Moreno and García-Caballero [44] provide a detailed account of Viète's proof in an appendix. Viète's product has been around for over 400 years: one might reasonably ask what new things can we learn from it. It turns out that we can learn a lot.

Recently mathematicians have returned to Viète's product. Levin (*op. cit.*) writes: "Given the simplicity, elegance, and age of Viète's product, it is surprising that there seems to have been few attempts at finding similar formulas." He recognized the crucial importance of the double angle formula $\cos 2z = 2\cos^2 z - 1$ in the proof of the product and generalized this structure to functions F that satisfy $F(\alpha z) = g(F(z))$ for some complex number α and some rational function g. This was a fruitful approach that led to a number of interesting results including Eq. (5.1.4) and another expression for square roots:

$$\sqrt{\frac{4 - z^2}{3}} = \sqrt{2 - z}\sqrt{2 - \sqrt{2 - z}}\sqrt{2 - \sqrt{2 - \sqrt{2 - z}}}\cdots, \qquad (5.2.1)$$

so that, for instance, with $z = 0$ we have

$$\frac{2}{\sqrt{3}} = \sqrt{2}\sqrt{2 - \sqrt{2}}\sqrt{2 - \sqrt{2 - \sqrt{2}}}\cdots.$$

Moreno and García-Caballero [44, 46] provide an alternative derivation of (5.2.1) and more "Viète-like" formulæ. In addition, various products have been derived connecting Lucas and Fibonacci numbers [24, 53, 54]. For example, Osler [53] showed that

$$
\frac{\sqrt{5}F_N}{2N\log\phi} = \sqrt{\frac{1}{2}+\frac{L_N}{4}}\sqrt{\frac{1}{2}+\frac{1}{2}\sqrt{\frac{1}{2}+\frac{L_N}{4}}}\sqrt{\frac{1}{2}+\frac{1}{2}\sqrt{\frac{1}{2}+\frac{1}{2}\sqrt{\frac{1}{2}+\frac{L_N}{4}}}}\cdots,
$$

where N is an even integer, $\phi = (1+\sqrt{5})/2$ is the golden ratio and F_N and L_N are Fibonacci and Lucas numbers respectively. (Osler gives a similar product for the case when N is an odd integer with $\sqrt{5}F_N$ and L_N interchanged.) Certainly a feature of the Viète-like products is the nested radicals. Osler [51] used nested radicals to derive another product representation for $\log x$.

The Wallis product for π has been known for some 350 years. Viète's product has the advantage of quicker convergence. In fact, Kreminski [34] has devised an algorithm so that Viète's product is even more efficient. The Wallis product, however, should be viewed outside the harsh light of numerical efficiency. A beautiful feature of this product is that it involves only *rational numbers* in the multiplication. The Viète and Wallis products represent the same number, but the products look very different. It is natural to ask whether there are further connections between these products. We know that both follow from infinite product expansions for $\sin z$, but are there more direct connections? Osler [52] showed a very simple and concrete connection. Define the sequence $\{V_M\}$ by $V_0 = 1$, and for $M \geq 1$,

$$
V_M = \prod_{n=1}^{M} a_n,
$$

where $a_1 = \sqrt{2}/2$ and $a_{n+1} = \sqrt{2+2a_n}/2$. The term V_M is simply the Mth partial product for the Viète product. Define the sequence $\{W_M\}$ by

$$
W_M = \prod_{n=1}^{\infty} \frac{2^{M+1}n-1}{2^{M+1}n} \cdot \frac{2^{M+1}n+1}{2^{M+1}n}.
$$

The W_M term is a depleted Wallis product, i.e. the Wallis product with the first M factors removed. For any non-negative integer M, Osler showed that

$$
\frac{2}{\pi} = V_M W_M. \tag{5.2.2}
$$

The Wallis product corresponds to the case $M = 0$, and Viète's product corresponds to the limiting case as $M \to \infty$. Equation (5.2.2) holds for any non-negative integer M and we can thus interpret the depleted Wallis product as the error factor when using partial products V_M to approximate $2/\pi$.

Wallis products were studied by Catalan (c. 1873), who showed that the product can be factored as follows:

$$\sqrt{2} = \frac{2\,2}{1\,3} \cdot \frac{6\,6}{5\,7} \cdot \frac{10\,10}{9\,11} \cdot \frac{14\,14}{13\,15} \cdots \tag{5.2.3}$$

$$\frac{\pi}{2\sqrt{2}} = \frac{4\,4}{3\,5} \cdot \frac{8\,8}{7\,9} \cdot \frac{12\,12}{11\,13} \cdot \frac{16\,16}{15\,17} \cdots . \tag{5.2.4}$$

Catalan also derived an infinite product for e,

$$e = \frac{2}{1} \left(\frac{4}{3}\right)^{1/2} \left(\frac{6\,8}{5\,7}\right)^{1/4} \left(\frac{10\,12\,14\,16}{9\,11\,13\,15}\right)^{1/8} \cdots , \tag{5.2.5}$$

and more recently Pippenger [55] derived the product

$$\frac{e}{2} = \left(\frac{2}{1}\right)^{1/2} \left(\frac{2\,4}{3\,3}\right)^{1/4} \left(\frac{4\,6}{5\,5} \cdot \frac{6\,8}{7\,7}\right)^{1/8} \cdots . \tag{5.2.6}$$

Sondow and Yi [63] summarize the above results and show that the two products for e are related through a simple identity. Though simply related, it is of interest to note that they were derived by very different means. Catalan used series and integral representations of the gamma function to get his product; Pippenger used only Stirling's approximation. Sondow and Yi (*op. cit.*) derive a number of other expressions for e and a generalization of the Wallis product:

$$\frac{\pi/M}{\sin(\pi/M)} = \frac{M}{M-1} \frac{M}{M+1} \cdot \frac{2M}{2M-1} \frac{2M}{2M+1} \cdot \frac{3M}{3M-1} \frac{3M}{3M+1} \cdots ,$$

for any integer $M \geq 2$. (The Wallis product is $M = 2$.) In fact, the product above was also derived (independently) by Moreno and García-Caballero [45] and used to good effect to give an infinite product representation for the golden ratio ϕ. The key observations are Eq. (3.2.1) and the fact that $\csc(\pi/10) = 2\phi$. They showed that

$$\phi = \frac{1}{2} \prod_{n=0}^{\infty} \frac{100n(n+1)+5^2}{100n(n+1)+3^2}.$$

There is, of course, a Viète-like expression for ϕ (cf. [25]).

Finally, we note that other constants of interest have infinite product representations. Melzak [43], for instance, showed that

$$\frac{\pi}{2e} = \prod_{n=1}^{\infty} \left(1 + \frac{2}{n}\right)^{(-1)^{n+1}n},$$

and derived a similar product for $6/(\pi e)$.

At the end of his paper on Viète products, Levin (*op. cit.*) writes: "One wonders what other constants have nice representations of this form. In any case, it has been shown that Viète's product is not an isolated, historical curiosity. The first analytic expression for π, though more than 400 years old, may still hold a few mysteries for us." The authors certainly agree.

Chapter 6
Tables of Products

An exhaustive collection of infinite product relations is not the purpose of these tables. The selection of what products to include is based mostly on two criteria: (a) the utility of the expression as measured by the use of it in elementary analysis; and (b) the sheer beauty of the relation. Some of the identities for constants reduce to simple judicious choices for variables in later identities. In these tables we summarize some of the product relations derived in the book, but we also give a number of other relations. The entries in the final column generally give the name (where known) of the mathematician, a reference (in bold type) if it is derived in this book and, often in lieu of a proof, references where the reader might find a proof of the result and generalizations. The products often depend on certain sequences and constants. We summarize these in the initial table.

© The Author(s), under exclusive license to Springer Nature Switzerland AG 2022
C. H. C. Little et al., *An Introduction to Infinite Products*, SUMS Readings,
https://doi.org/10.1007/978-3-030-90646-7_6

Constants and Sequences

B_n	Bernoulli	$$\frac{z}{e^z - 1} = \sum_{n=0}^{\infty} B_n \frac{z^n}{n!},$$ $$|z| < 2\pi$$	$B_{2n+1} = 0, \quad n \geq 1$ $B_0 = 1, \quad B_1 = -\frac{1}{2}, \quad B_2 = \frac{1}{6},$ $B_4 = -\frac{1}{30}, \quad B_6 = \frac{1}{42},$ $B_8 = -\frac{1}{30}, \quad \cdots$
F_n	Fibonacci	$F_0 = 0, \quad F_1 = 1$ $F_n = F_{n-1} + F_{n-2}$	$0, 1, 1, 2, 3, 5, 8, 13, 21, 34, \ldots$
L_n	Lucas	$L_0 = 2, \quad L_1 = 1$ $L_n = L_{n-1} + L_{n-2}$	$2, 1, 3, 4, 7, 11, 18, 29, 47, 76, \ldots$
G_n		$$G_n = \prod_{k=0}^{n} (k+1)^{(-1)^{k+1}\binom{n}{k}}$$	$G_1 = \frac{2}{1}, \quad G_2 = \frac{2^2}{1 \cdot 3},$ $G_3 = \frac{2^3 \cdot 4}{1 \cdot 3^3}, \quad G_4 = \frac{2^4 \cdot 4^4}{1 \cdot 3^6 \cdot 5}, \cdots$
p_n	prime numbers	$p_1 = 2,$ $p_n = n^{th} \text{prime}$	$2, 3, 5, 7, 11, 13, 17, 19, 23, \ldots$
ϕ	golden ratio	$$\phi = \frac{1 + \sqrt{5}}{2}$$	≈ 1.6180
γ	Euler's constant	$$\gamma = \lim_{n \to \infty} \left(\sum_{k=1}^{n} \frac{1}{k} - \log n \right)$$	≈ 0.5772

I Products for π

1	$$\frac{2}{\pi} = \prod_{n=1}^{\infty} a_n$$	$a_1 = \dfrac{\sqrt{2}}{2},$ $a_{n+1} = \dfrac{\sqrt{2 + 2a_n}}{2}$	Viète **(2.5.7)** [44] (translation of Viète's proof)
2	$$\frac{\pi}{2} = \prod_{n=1}^{\infty} \frac{4n^2}{(2n-1)(2n+1)}$$		Wallis **(2.6.8)**
3	$$\frac{\pi}{2} = \prod_{n=1}^{\infty} (G_n)^{1/2^n}$$		Sondow [64]
4	$$\frac{\pi^2}{6} = \prod_{n=1}^{\infty} \frac{p_n^2}{p_n^2 - 1}$$		Euler (n=1 in VI-5)

II Products for e

1	$$e = \frac{2}{1} \left(\frac{4}{3}\right)^{1/2} \left(\frac{6}{5}\frac{8}{7}\right)^{1/4} \left(\frac{10}{9}\frac{12}{11}\frac{14}{13}\frac{16}{15}\right)^{1/8} \cdots$$		Catalan [63]
2	$$\frac{e}{2} = \left(\frac{2}{1}\right)^{1/2} \left(\frac{2}{3}\frac{4}{3}\right)^{1/4} \left(\frac{4}{5}\frac{6}{5}\frac{6}{7}\frac{8}{7}\right)^{1/8} \cdots$$		Pippenger [55], [63]
3	$$e = \prod_{n=1}^{\infty} (G_n)^{1/n}$$		Guillera [64] (x=1 in IV-8)
4	$$e = \prod_{n=1}^{\infty} e_n$$	$e_1 = 1,$ $e_{n+1} = (n+1)(e_n + 1)$	[38]

	III Products for Miscellaneous Constants		
1	$\dfrac{\pi}{2e} = \displaystyle\prod_{n=1}^{\infty}\left(1+\dfrac{2}{n}\right)^{(-1)^{n+1}n}$		Melzak [43]
2	$\dfrac{6}{\pi e} = \displaystyle\prod_{n=2}^{\infty}\left(1+\dfrac{2}{n}\right)^{(-1)^{n}n}$		Melzak [43]
3	$e^{\gamma} = \displaystyle\prod_{n=1}^{\infty}(G_n)^{1/(n+1)}$		Sondow [64]
4	$\log 2 = \displaystyle\prod_{n=1}^{\infty}\dfrac{2}{1+2^{1/2^n}}$		von Seidel (x=2 in IV-9)
5	$\sqrt{2} = \displaystyle\prod_{n=1}^{\infty}\left(1+\dfrac{1}{q_n}\right)$	$q_1 = 3,$ $q_{n+1} = 2q_n^2 - 1$	Cantor (q_1=3 in IV-1)
6	$\sqrt{2} = \dfrac{2}{1}\dfrac{2}{3}\cdot\dfrac{6}{5}\dfrac{6}{7}\cdot\dfrac{10}{9}\dfrac{10}{11}\cdot\dfrac{14}{13}\dfrac{14}{15}\cdots$		Catalan [63]
7	$\phi = \dfrac{1}{2}\displaystyle\prod_{n=0}^{\infty}\dfrac{100n(n+1)+5^2}{100n(n+1)+3^2}$		Moreno [45]
8	$\dfrac{3}{\phi} = \displaystyle\prod_{n=1}^{\infty}\left(1+\dfrac{1}{F_{2^n+1}}\right)$		Sondow [65]
9	$3-\phi = \displaystyle\prod_{n=1}^{\infty}\left(1+\dfrac{1}{L_{2^n+1}}\right)$		Sondow [65]

IV Elementary Functions

1	$$\sqrt{\frac{x+1}{x-1}} = \prod_{n=1}^{\infty}\left(1+\frac{1}{q_n}\right)$$	$q_1 = x > 1,$ $q_{n+1} = 2q_n^2 - 1$	Cantor [22]
2	$$\sin z = z\prod_{n=1}^{\infty}\left(1-\frac{z^2}{n^2\pi^2}\right)$$	$z \in \mathbb{C}$	Euler **Theorem 2.6.1**
3	$$\sin z = \frac{2z}{\pi}\prod_{n=1}^{\infty}\left(1+\frac{4z^2-\pi^2}{(1-4n^2)\pi^2}\right)$$	$z \in \mathbb{C}$	Melnikov [41]
4	$$\sin z = z\prod_{n=1}^{\infty}\cos\left(\frac{z}{2^n}\right)$$	$z \in \mathbb{C}$	Euler **(2.5.6)**
5	$$\frac{\sin\alpha z}{\sin\beta z} = \frac{\alpha}{\beta}\prod_{n=1}^{\infty}\left(1+\frac{(\alpha^2-\beta^2)z^2}{\beta^2 z^2 - n^2\pi^2}\right)$$	α,β constants $z \in \mathbb{C},$ $\beta z \neq m\pi, m \in \mathbb{Z}$	Melnikov [41]
6	$$\cos z = \prod_{n=1}^{\infty}\left(1-\frac{4z^2}{(2n-1)^2\pi^2}\right)$$	$z \in \mathbb{C}$	Euler **Theorem 2.6.1**
7	$$2^z = \prod_{n=1}^{\infty}\left(1-\frac{(-1)^n z}{n-z}\right)$$	$z \in \mathbb{C},$ $z \neq 1,2,3\ldots$	**(3.5.9)**
8	$$e^x = \prod_{n=1}^{\infty}\left(\prod_{k=0}^{n}(kx+1)^{(-1)^{k+1}\binom{n}{k}}\right)^{1/n}$$	$x > 0$	Guillera, Sondow [66]
9	$$\frac{\log x}{x-1} = \prod_{n=1}^{\infty}\frac{2}{1+x^{1/2^n}}$$	$x > 1$	von Seidel [38], [35]

	V Products Involving $\Gamma(z)$		
1	$$\frac{1}{\Gamma(z)} = ze^{\gamma z} \prod_{n=1}^{\infty} \left(1 + \frac{z}{n}\right) e^{-z/n}$$	$z \in \mathbb{C}$	Euler **(3.1.3)**
2	$$\Gamma(z) = \frac{1}{z} \prod_{n=1}^{\infty} \frac{\left(1 + \frac{1}{n}\right)^z}{1 + \frac{z}{n}}$$	$z \in \mathbb{C}$, $z \neq 0, -1, -2, \ldots$	Euler **(3.1.11)**
3	$$\prod_{n=1}^{\infty} \frac{(n - a_1)(n - a_2)}{(n - b_1)(n - b_2)}$$ $$= \frac{\Gamma(1 - b_1)\Gamma(1 - b_2)}{\Gamma(1 - a_1)\Gamma(1 - a_2)}$$	$a_1 + a_2 = b_1 + b_2$ $a_k, b_k \neq 1, 2, 3 \ldots$	**(3.5.3)**
4	$$\prod_{n=0}^{\infty} \left(1 - \frac{x^2}{(n + y)^2}\right)$$ $$= \frac{\Gamma^2(y)}{\Gamma(y + x)\Gamma(y - x)}$$	$y \, , y + x, y - x$ $\neq 0, -1, -2, \ldots$	**Example 3.5.1**

VI Products Involving $\zeta(s)$, $s > 1$

1	$$\zeta(s) = \prod_{n=1}^{\infty} \frac{p_n^s}{p_n^s - 1}$$ $$= \sum_{n=1}^{\infty} \frac{1}{n^s}$$		Euler **(4.1.14)**
2	$$\frac{\zeta^2(s)}{\zeta(2s)} = \prod_{n=1}^{\infty} \frac{p_n^s + 1}{p_n^s - 1}$$ $$= \sum_{n=1}^{\infty} \frac{2^{\omega(n)}}{n^s}$$	$\omega(n)$ is the number of different prime factors of n	[30], p. 255
3	$$\frac{\zeta(2s)}{\zeta(s)} = \prod_{n=1}^{\infty} \frac{p_n^s}{p_n^s + 1}$$ $$= \sum_{n=1}^{\infty} \frac{(-1)^{\rho(n)}}{n^s}$$	$\rho(n)$ is the total number of prime factors of n	[30], p. 255
4	$$\frac{\zeta^4(s)}{\zeta(2s)} = \prod_{n=1}^{\infty} \frac{p_n^{2s}(p_n^s + 1)}{(p_n^s - 1)^3}$$ $$= \sum_{n=1}^{\infty} \frac{d^2(n)}{n^s}$$	$d(n)$ is the number of divisors of n including 1 and n	Ramanujan [30], p. 256
5	$$\zeta(2n) = \prod_{n=1}^{\infty} \frac{p_n^{2n}}{(p_n^{2n} - 1)}$$ $$= \frac{(-1)^{n+1} 2^{2n-1} \pi^{2n} B_{2n}}{(2n)!}$$	$n \in \mathbb{N}$	**(2.6.17)**

VII Products and Series from Partitions

1	$$\prod_{n=1}^{\infty}\frac{1}{1-x^n}=1+\sum_{n=1}^{\infty}p(n)x^n$$	$\lvert x\rvert<1$ $p(n)$ is the number of partitions of n	Euler **Theorem 4.2.4**
2	$$\prod_{n=1}^{\infty}(1-x^n)$$ $$=1+\sum_{n=1}^{\infty}(-1)^n\left(x^{g(n)}+x^{g(-n)}\right)$$	$\lvert x\rvert<1$ $g(n)=\dfrac{n(3n-1)}{2}$	Euler **Corollary 4.3.6**
3	$$\prod_{n=1}^{\infty}\frac{1-x^{2n}}{1-x^{2n-1}}=\sum_{n=0}^{\infty}x^{n(n+1)/2}$$	$\lvert x\rvert<1$	Gauss **Corollary 4.3.8**
4	$$\prod_{n=0}^{\infty}\frac{1+\alpha x q^n}{1-x q^n}$$ $$=1+\sum_{n=1}^{\infty}\prod_{k=1}^{n}\left(\frac{1+\alpha q^{k-1}}{1-q^k}\right)x^n$$	$\lvert x\rvert<1$ $\lvert q\rvert<1$ $\alpha=\text{const.}$	Cauchy **Theorem 4.3.2**
5	$$\prod_{n=0}^{\infty}(1+xq^n)$$ $$=1+\sum_{n=1}^{\infty}\frac{q^{n(n-1)/2}x^n}{\prod_{k=1}^{n}(1-q^k)}$$	$\lvert q\rvert<1$	Euler **Corollary 4.3.4**
6	$$\prod_{n=0}^{\infty}\frac{1}{1-xq^n}$$ $$=1+\sum_{n=1}^{\infty}\frac{x^n}{\prod_{k=1}^{n}(1-q^k)}$$	$\lvert x\rvert<1$ $\lvert q\rvert<1$	Euler **Corollary 4.3.4**
6	$$\prod_{n=0}^{\infty}\left(1-q^{2n+2}\right)\left(1+xq^{2n+1}\right)$$ $$\times\left(1+x^{-1}q^{2n+1}\right)$$ $$=\sum_{n=-\infty}^{\infty}q^{n^2}x^n$$	$x\neq0$ $\lvert q\rvert<1$	Jacobi **Theorem 4.3.5**

Bibliography

1. Abramowitz, M, Stegun, I., *Handbook of Mathematical Functions*, Dover, 1965.
2. Ahlfors, L., *Complex Analysis*, McGraw-Hill, 2nd ed., 1966.
3. Allouche, J.-P., "Paperfolding infinite products and the gamma function", *J. Num. Th.*, vol. **148**, pp. 95–111, 2015.
4. Almodovar, L., Moll, V.H., Quan, H., Roman, F., Rowland, E., Washington, M., "Infinite products arising in paperfolding", *J. Int. Seq.*, vol. **19** Art no 16 5 1, 2016.
5. Andrews, G.E., *The Theory of Partitions*, Cambridge University Press, 1998.
6. Andrews, G.E., *Number Theory*, Saunders, 1971.
7. Apostol, T.M., *Mathematical Analysis*, Addison-Wesley, 2nd ed., 1974.
8. Apostol, T.M., *Introduction to Analytic Number Theory*, Springer-Verlag, 1976.
9. Artin, E., *The Gamma Function*, Holt, Rinehart and Winston, 1964.
10. Ash, T.B., Novinger, W.V., *Complex Variables*, Dover, 2004.
11. Bateman, P.T., Diamond, H.G., "A hundred years of prime numbers", *Amer. Math. Monthly*, vol. **103**, pp. 729–41, 1996.
12. Boas, R.J., *Entire Functions*, Academic Press, 1954.
13. Brock, J.E., "An example in double series", *Amer. Math. Monthly*, vol. **50**, p. 619, 1943.
14. Bromwich, T.J.I'A., *An Introduction to the Theory of Infinite Series*, Macmillan, 1926.
15. Brown, J.W. and Churchill, R.V., *Complex Variables and Applications*, McGraw-Hill, 7th ed., 2004.
16. Chamberland, M., Straub, A., "On gamma quotients and infinite products", *Adv. Appl. Math.*, vol. **51**, pp. 546–62, 2013.
17. Chaudhry, M.A., Qadir, A., Rafique, M., Zubair, S.M., "Extension of Euler's beta function", *J. Comp. & Appl. Math.*, vol. **78**, pp. 19–32, 1997.
18. Colwell, P. "On the boundary behavior of Blaschke products in the unit disk", *Proc. AMS*, pp. 582–7, 1966.
19. Colwell, P. *Blaschke Products—Bounded Analytic Functions*, University of Michigan Press, 2016.
20. Davis, P.J., "Leonhard Euler's integral: a historical profile of the gamma function", *Amer. Math. Monthly*, vol. **66**, pp. 849–69, 1959.
21. Fine, N.J., *Basic hypergeometric series and applications*, Mathematical Surveys and Monographs, No. 27, American Mathematical Society, 1988.
22. Fine, N.J., "Infinite Products for k-th Roots", *Amer. Math. Monthly*, vol. **84**, no 8, pp. 629–30, 1977.
23. Fulks, W., *Advanced Calculus*, 3rd ed., Wiley, 1978.

24. García-Caballero, E.M., Moreno, S.G., Prophet, M.P., "A complete view of Viète-like infinite products with Fibonacci and Lucas numbers", *Appl. Math. Comp.*, vol. **247**, pp. 703–11, 2014.
25. García-Caballero, E.M., Moreno, S.G., Prophet, M.P., "The golden ratio and Viète's formula", *Teaching Math. and Comp. Sci.*, vol. **12** no.1, pp. 43–54, 2014.
26. Gasper, G., Rahman, M., *Basic Hypergeometric Series*, Cambridge University Press, 1990.
27. Gonzàlez, M., *Complex Analysis: Selected Topics*, Marcel Dekker, 1992.
28. Gradshteyn, I.S., Ryzhik, I.M., *Table of Integrals, Series, and Products*, 4th ed., Academic Press, 1980.
29. Hardy, G.H., "A note on the continuity or discontinuity of a function defined by an infinite product", *Proc. Lond. Math. Soc.*, vol. **7**, pp. 40–8, 1908.
30. Hardy, G.H., Wright, E.M., *An Introduction to the Theory of Numbers*, Oxford University Press, 5th ed., 1979.
31. Hochstadt, H., *Integral Equations*, Wiley, 1973.
32. Ingham, A.E., *The Distribution of Prime Numbers*, Cambridge University Press, 1990.
33. Jameson, G.J.O., "A fresh look at Euler's limit formula for the gamma function", *Math. Gazette*, vol. **98**, pp. 235–42, 2014.
34. Kreminski, R., "π to thousands of digits from Vieta's formula", *Mathematics Magazine*, vol. **81**, no 3, pp.201–7, 2008.
35. Levin, A., "A new class of infinite products generalizing Viète's product formula for π", *Ramanujan J.*, vol. **10**, pp. 305–24, 2005.
36. Little, C.H.C., Teo, K.L., van Brunt, B., *Real Analysis via Sequences and Series*, Springer-Verlag, 2015.
37. Littlewood, J.E., "On a class of conditionally convergent infinite products", *Proc. Lond. Math. Soc.*, vol. **8**, pp. 195–9, 1910.
38. Loya, P., *Amazing and Aesthetic Aspects of Analysis*, Springer, 2017.
39. Marsden, J.E., Hoffman, M.J., *Basic Complex Analysis*, 2nd ed., Freeman, 1987.
40. Mashreghi, J., Fricain, E. (eds.), *Blaschke Products and their Applications*, Fields Institute Communications, Springer, 2013.
41. Melnikov, Y.A., "A new approach to the representation of trigonometric functions and hyperbolic functions by infinite products", *J. Math. Anal. App.*, vol. **344**, pp. 521–34, 2008.
42. Melnikov, Y.A., *Green's Functions and Infinite Products*, Birkhäuser, 2011.
43. Melzak, Z.A., "Infinite products for πe and π/e", *Amer. Math. Monthly*, vol. **68**, pp. 39–51, 1961.
44. Moreno, S.G., García-Caballero, E.M., "On Viète-like formulas", *J. Approx. Theory*, vol. **174**, pp. 90–112, 2013.
45. Moreno, S.G., García-Caballero, E.M., "An infinite product for the golden ratio", *Amer. Math. Monthly*, vol. **119**, no 8, p. 645, 2012.
46. Moreno, S.G., García-Caballero, E.M., "Infinite products of cosines and Viète-like formulæ", *Math. Magazine*, vol. **86**, no 1, pp. 15–25, 2013.
47. Moritz, R.E., "On the extended form of Cauchy's condensation test for the convergence of infinite series", *Bull. Amer. Math. Soc.*, vol. **44**, pp. 441–2, 1938.
48. Nimbran, A.S., "An infinite product for the square root of an integer", *Amer. Math. Monthly*, vol. **124**, no 7, pp. 647–50, 2017.
49. Osler, T.J., "An Unusual Product for $\sin z$ and variations of Wallis's product", *Math. Gazette*, vol. **87**, no 508, pp. 134–9, 2003.
50. Osler, T.J., "Interesting finite and infinite products from simple algebraic identities", *Math. Gazette*, vol. **90**, no 517, pp. 90–3, 2006.
51. Osler, T.J., Jacob, W. & Nishimura, R., "An infinite product of nested radicals for log x from the Archimedean algorithm", *Math. Gazette*, vol. **100**, no 548, pp. 274–8, 2016.
52. Osler, T.J., "The union of Vieta's and Wallis's products for Pi", *Amer. Math. Monthly*, vol. **106**, no 8, pp. 774–6, 1999.
53. Osler, T.J., "Vieta-like products of nested radicals", *Math. Gazette*, vol. **94**, no 529, pp. 62–6, 2010.

54. Osler, T.J., "Vieta-like products of nested radicals with Fibonacci numbers and Lucas numbers", *Fibonacci Quart.*, vol. **45**, pp.202–4, 2008.
55. Pippenger, N., "An infinite product for e", *Amer. Math. Monthly*, vol. **87**, p. 391, 1980.
56. Priestley, H.A., *Introduction to Complex Analysis*, Rev. ed., Oxford University Press, 1990.
57. Pustylnik, E., Reich, S., "Infinite products of arbitrary operators and intersections of subspaces in Hilbert space", *J. Approx. Theory*, vol. **178**, pp. 91–102, 2014.
58. Ramasinghe, W., "Inspiring examples in rearrangements of infinite products", *Int. J. Math. Ed. in Science and Technology*, vol. **38**, pp. 797–804, 2007.
59. Reich, S., Salinas, Z., "Weak convergence of infinite products of operators in Hadamard spaces", *Read. Circ. Mat. Palermo*, vol. **65**, pp. 51–71, 2016.
60. Reich, S., Zaslavski, A.J., *Genericity in Nonlinear Analysis*, Springer-Verlag, 2014.
61. Remmert, R., "Wielandt's theorem about the Γ-function", *Amer. Math. Monthly*, vol. **103**, pp. 214–20, 1996.
62. Segal, S.L., *Nine Introductions in Complex Analysis*, rev. ed., North-Holland, 2007.
63. Sondow, J., Yi HH., "New Wallis and Catalan type infinite products for α, e and $\sqrt{2 + \sqrt{2}}$", *Amer. Math. Monthly*, vol. **117**, pp. 912–7, 2010.
64. Sondow, J.,"A faster product for π and a new integral for $\ln \pi/2$", *Amer. Math. Monthly*, vol. **112**, pp. 729–34, 2005.
65. Sondow, J., "Evaluation of Tachiya's algebraic infinite products involving Fibonacci and Lucas numbers", *DARF 2011, Conf. Proc. Sept. 6, 2011*, **1385**, pp. 97–100, 2011.
66. Sondow, J., Guillera, J., "An infinite product for the exponential", Problem 11381, *Amer. Math. Monthly*, (proposal) vol. **115**, p. 665, 2008; (solution) vol. **117**, pp. 283–4, 2010.
67. Titchmarsh, E.C., *The Theory of Functions*, 2nd ed., Oxford University Press, 1939.
68. Venkatachaliengar, K., "Elementary proofs of the infinite product for $\sin z$ and allied formulæ", *Amer. Math. Monthly*, vol. **69**, pp. 541–5, 1962.
69. Whittaker, E.T., Watson, G.N., *A Course of Modern Analysis*, 4th ed., Cambridge University Press, 1952.
70. Wingler, E., "An Infinite product expansion for the square root function", *Amer. Math. Monthly*, vol. **97**, no 9, pp. 836–9, 1990.
71. Zagier, D., "Newman's short proof of the prime number theorem", *Amer. Math. Monthly*, vol. **104**, pp. 705–8, 1997.
72. Zhang, G.-H., *Theory of Entire and Meromorphic Functions: Deficient and Asymptotic Values and Singular Directions*, Trans. of Math. Mono. vol. **122**, American Math. Soc., 1991.

Index

Printed in the United States
by Baker & Taylor Publisher Services